Kister Distillation Symposium 2021

Topical Conference at the 2021 AIChE Spring Meeting and 17th Global Congress on Process Safety

Online
18 - 23 April 2021

ISBN: 978-1-7138-5333-6

Kister Distillation Symposium 2021

Topical Conference at the 2021 AIChE Spring Meeting and 17th Global Congress on Process Safety

Online
18 - 23 April 2021

AIChE
The Global Home of Chemical Engineers

Printed from e-media with permission by:

Curran Associates, Inc.
57 Morehouse Lane
Red Hook, NY 12571

Some format issues inherent in the e-media version may also appear in this print version.

AIChE
120 Wall Street, FL 23
New York, NY 10005-4020

Phone: (800) 242-4363
Fax: (203) 775-5177

www.aiche.org

Additional copies of this publication are available from:

Curran Associates, Inc.
57 Morehouse Lane
Red Hook, NY 12571 USA
Phone: 845-758-0400
Fax: 845-758-2633
Email: curran@proceedings.com
Web: www.proceedings.com

TABLE OF CONTENTS

618132 Mega Columns: Challenges and Solutions

Abstract:

As processes become more complex and economies of scale drive plant economics, mega columns become more and more common. This presentation discusses challenges associated with the design of mega columns and presents innovative solutions to those new challenges.

Examples of processes requiring extremely tall columns and/or extremely large diameters will be discussed. With these columns, there are unique considerations for hydraulic feed and distribution, the optimal type of column internals, mechanical support of internals, achievable and allowable mechanical design tolerances, as well as installation and revamp logistics. The sheer size of such processes and columns creates huge economic pressures. Correct design, efficient startup, and proper operation of these columns is imperative.

618658 Reuse and Relocation: A Case History Part I Reboiler Analysis and Performance

Abstract:

Reuse and Relocation: A Case History Part I

Reboiler Analysis and Performance

Andrew W. Sloley

Potentially, used equipment can save time and money over new equipment in your plant. Occasionally, used equipment can be picked up for very low prices. But getting it to the plant and properly working does require skills that differ from purchasing new equipment.

However, the process requirements for the new service are nearly always different from the original process requirements. Expert-level knowledge may be required to decide if the equipment will really do the required job.

Unless an entire plant is purchased and used as-is, used equipment will have to be integrated with other parts of the plant that will use new equipment. Integration may be difficult as the used equipment may have constraints that makes its use difficult with conventional choices for the new equipment attached to it.

This presentation examines a plant that purchased a 42-inch diameter tower used from a pharmaceutical plant. It had been built and erected but never placed into service. The client purchased the tower and placed it into a completely new service.

The focus on this presentation is on the reboiler system and difficulties with this system. The reboiler chosen was a falling-film evaporator. Once in service, the reboiler could not get the required duty into the tower. Areas covered include:

- Investigating leaks in the reboiler,
- Determining the actual reboiler duty,
- Examining the actual capability of the hot oil system providing heat to the reboiler,
- Inherent limits on the reboiler performance including fouling and liquid distribution problems,
- Steps taken to improve reboiler performance, and
- The alternate best case reboiler type that would have performed better.

619619 The Lost Art of Tower Internal Inspection – A Practical Guide for the New Generation

Abstract:

Although a maintenance outage in an oil refinery may be viewed as a period of downtime in revenue earning and increased expense, in reality, this is the stage through which proper planning and inspection can have the potential for increased revenue, improved safety, and a reduction in operating expense for the next run cycle. During a maintenance outage, the Process Engineer is provided with the opportunity and freedom to crawl inside the equipment. With this freedom comes great responsibility. If it can affect the performance of the process, it falls on the shoulders of the Process Engineer. It takes significant preparation to perform the task of tower internal inspection accurately and efficiently. This preparation involves everything from understanding the operational objectives and constraints of the equipment prior to the shutdown, to studying vessel and tray drawings. This preparation also includes reviewing recent vessel gamma scans and having the equipment modeled in a process simulation ahead of time. There is a level of physical and mental preparation that is needed in addition to possessing essential tools. This paper will serve as practical guide for Process Engineers in preparing for and conducting a successful tower internal inspection. It will also empower them to effectively communicate with the Maintenance Organization and Management to positively influence change as it relates to improving the performance of the process via the findings of their inspection; thus significantly contributing to their company's bottom line for years to come.

619618 Is Larger-Scale Distillation Research Still Relevant?

Abstract:

In the era of quantum computing, nanotechnology, and artificial intelligence, is the large-scale distillation research still relevant? The liquid/vapor distribution, downcomer hydraulics, flow-pass length and bed-length effect, concentration gradient etc. are very crucial to the performance of column internals. However, all these phenomena cannot be studied in small laboratory columns due to the significant difference in the 'shape of one theoretical stage' between laboratory and industrial scale distillation research. The role of the large-scale distillation research and challenges of conducting the research will be summarized. Opportunities of conducting large-scale heat transfer studies will also be presented. The future of the large-scale distillation research will be briefly discussed.

618726 Debottleneck Distillation Columns Using Dividing Wall Column Technology

Abstract:

Dividing wall technology is increasingly used in the industry to produce more throughput in existing columns.

This presentation explores the revamp journey of two commonly used separation sequences:

Case study 1: A sequence of Naphtha Splitters were revamped multiple times over its 15 years of life cycle. The original configuration of columns was equipped with conventional sieve trays. In the 1st revamp the column internals were changed to high capacity packing. Though the capacity increase was more than 40%, the product specifications were compromised.

In the 2nd revamp, an additional side cut was introduced to improve product specifications. The drawback was a reduction of top and bottom product yields.

Finally, the column was revamped to a dividing wall column. **The revamp provided a capacity increase of an additional 40% over that achieved by high capacity internals**. There was simultaneous improvement in the product specifications and yields.

Case Study 2: This case study involved an indirect sequence of a depropanizer followed by propylene splitter producing chemical grade propylene. The two columns were equipped with high capacity trays. Besides a 30% increase in through put, DWC revamp provided higher propylene recovery and lower RVP of bottoms C4 product used for gasoline blending.

619628 Reducing FCC Main Fractionator Operating Risks

Abstract:

Light Tight Oil (LTO) crude processing in US and foreign refineries has been substantially increased over the past 10 years. LTO is highly paraffinic and has a very high percentage of Kerosene and lighter boiling range material yield per barrel of crude. Domestic production of LTO is very attractive to US refineries as it reduces transportation costs substantially. LTO processing experience is increasing after many processing learning lessons, mainly in blending and Crude Distillation Unit (CDU) and Vacuum Distillation Unit (VDU), have been cited in industry publications.

CDU and VDUs are not the only units that have experienced operating issues. LTO Atmospheric Gas Oil (AGO), Vacuum Gas Oil (VGO) and Atmospheric Tower Bottoms (ATB) are excellent feedstocks for the Fluid Catalytic Cracker (FCC) unit due to its paraffinicity and low metals / con-carbon content. Those feedstocks result in high conversion and excellent product slates, without high catalyst contamination. Excellent catalytic conversion of this feedstock sometimes makes the FCC unit operations challenged because adequate main fractionator operation and run-down goals should be maintained with minimum Slurry (FCC Main Fractionator Bottoms) yield. At high FCC reactor temperature, the LTO feedstock will approach terminal conversion, which places particular challenges on the operation with this LTO feedstock.

Many FCC main fractionator designs are heritages from previous refinery operating strategies. These fractionator designs have sections of the column that pose large operating risks without providing measurable benefits. In the current market, it is important that operating risks be minimized in order to maximize feedstock flexibility while minimizing shutdown risks.

In this case study, the authors will discuss a recent a recent FCC troubleshooting and evaluation. The FCC main fractionator was a traditional design with a Heavy Cycle Oil (HCO) pumparound, HCO wash, and Slurry pumparound sections in the lower half of the fractionator. Successful HCO wash section operation became challenging due to the heat removal requirements while processing LTO feedstocks. HCO wash section performance evaluation showed that the HCO wash section removal had little- to no-effect on the overall column operation. In addition, impacts on the Light Cycle Oil (LCO) / Slurry fractionation section were minimal in order to fully compensate for the impacts of losing the HCO wash section (HCO / Slurry fractionation). Eliminating the HCO wash section reduces the fractionator operating risk as this section has a very high vapor rate, low liquid flux, and very high temperatures which make it prone to coking.

AIChE Paper Number: 98c

Destructive Examination of Ultra-Selective Exchangers (USX's)

Jonathan D. McFarlen
Consulting Engineer
M&M Engineering,
a Division of Acuren Inspection, Inc.

Thomas Garza, PE
Manager - Fixed Equipment, QA/QC, and Inspection
Chevron Phillips Chemical Company LP

Prepared for Presentation at the 2021 Spring National Meeting
Virtual, April 18 – 23, 2021

Destructive Examination of Ultra-Selective Exchangers (USX's)

Jonathan D. McFarlen
Consulting Engineer
M&M Engineering,
a Division of Acuren Inspection, Inc.

Thomas Garza, PE
Manager - Fixed Equipment, QA/QC and Inspection
Chevron Phillips Chemical Company LP

Abstract

Chevron Phillips Chemical Company LP ("CPChem") operates Stone & Webster cracking furnaces at its various ethylene plants in the Gulf Coast area. Stone & Webster furnace technology utilizes BORSIG USX-Omega primary quench heat exchangers. The USX exchangers are designed to quench the heater effluent by producing steam. Boiler feedwater (BFW) flows from the steam drum through the outer tube of the vertical USX exchangers. Each USX has a primary and secondary pass. During a process upset, the USX exchangers were exposed to temperatures that exceeded their design conditions.

A Fitness-for-Service (FFS) assessment was commissioned on the USX-Omega exchangers after the equipment operated above the design temperature limits for approximately 5 hours. The FFS assessment followed API 579-1/ASME FFS-1, Part 10 methodology (Assessment of Components Operating in the Creep Range), starting with a Level 3 engineering analysis (Phase 1) followed by field metallurgical evaluation in conjunction with non-destructive testing (Phase 2), and concluded with destructive analysis of exchanger samples (Phase 3).

This paper discusses API 579-1/ASME FFS-1 Part 10, Assessment of Components Operating in the Creep Range, destructive metallurgical testing that was pursued to establish suitable remaining life so that the equipment could be returned back to service with assurance of safe, reliable operation.

Introduction and Background

Chevron Phillips Chemical Company LP ("CPChem") operates Stone & Webster cracking furnaces at its various ethylene plants in the Gulf Coast area. In March 2019, the design temperatures of the radiant coil's outlets, USX-Omega's inlets and outlets and the inlet to the Secondary Transfer Line Exchanger (STLE) were exceeded due to a possible swing in the fuel gas composition. The USX-Omega exchangers inlet and outlet design temperatures of 1750°F and 750°F, respectively, were exceeded for a period of approximately 5 hours. Due to the process event that was experienced by the USX-Omega's there was concern that the heat exchanger materials had suffered thermal and possibly creep damage.

Knighthawk Engineering, Inc. (KHE) was contracted to perform a Fitness-for-Service (FFS) assessment of the USX-Omega's in accordance with the API-579/ASME FFS-1 Code, Part 10, Assessment of Components Operating in the Creep Range, starting with a Level 3 engineering analysis. The Level 3 Assessment performed by KHE included process data analysis, process analysis, heat transfer analysis, finite element analysis, materials testing and metallurgical analysis. The Finite Element Analysis (FEA) thermal-mechanical analysis performed by KHE indicated that the USX-Omega's primary legs and return bends did not meet the requirements of the API 579/ASME FFS-1, Part 10, Level 3 analysis because localized plastic strain at the inner pipe to header box weld area exceeding 2.5% (the strain limit at welds).

As permitted by API 579-1/ASME FFS-1 Part 10, Assessment of Components Operating in the Creep Range, non-destructive testing, and destructive metallurgical testing was pursued to qualify the equipment/components for continued service. Creep rate testing was part of the test plan due to temperature limitations that defines the creep range of 650°F for carbon steel and 750°F for carbon – ½ molybdenum steel (ref. API 579-1/ASME FFS-1, Table 4.1). M&M Engineering, a Division of Acuren Inspection, Inc (Acuren/MME) was contracted to perform metallurgical examination (both non-destructive and destructive) to determine if incipient creep damage or thermal alteration had occurred to the USX-Omega exchangers during the temperature excursion.

The exchangers that were evaluated are BORSIG USX-Omega Quench Exchangers which utilize a double pipe design that is close coupled to the outlet of the radiant coil. The USX-Omega exchangers incorporate inlet and outlet turbo chambers which eliminates the possibility of solid build-up due to the tangential design of the inlet and outlet nozzles. This particular design utilizes three different types of refractory in the annulus of inlet nozzle, which improves the thermal gradient uniformly along the length of the nozzle. The main benefit of the USX-Omega is that the cracked gas from the radiant outlet coil is individually quenched in its own double pipe element. **Figure 1** shows an overview of the USX-Omega and its various components.

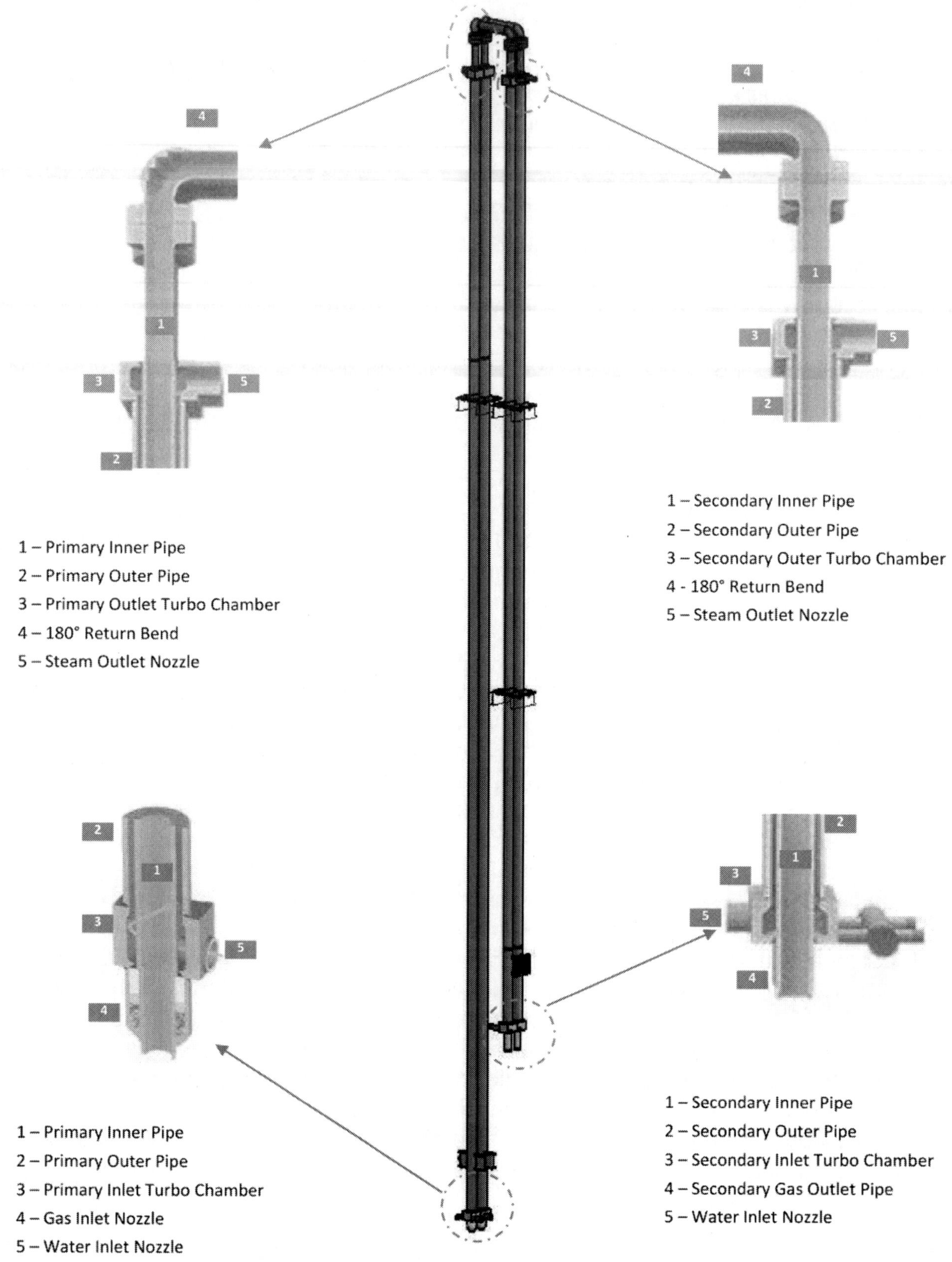

Courtesy of BORSIG Process Heat Exchanger GmbH

Figure 1 – Typical BORSIG USX-Omega

Heat Exchanger Data

The USX-Omega exchangers were designed for a maximum gas inlet (nozzle) temperature of 1751°F and 698°F maximum design temperature for the primary and secondary legs inner pipes (which are water/steam jacketed).

Materials of Construction of key components and Codes of construction used for the FFS evaluation are as follows:

Component	Material of Construction
USX Inner Pipes (Primary/Secondary)	A335 Grade P1 (C – ½ Moly)
USX Outer Pipes (Primary/Secondary)	ASTM A106 Gr. B
USX Gas Inlet Nozzles	ASTM B408, Incoloy™ 800HT
USX Header Boxes (Primary/Secondary)	ASTM A516 Gr. 70
USX Primary Outlet Connecting Pipe	ASTM A312 Type 321
USX Secondary Legs Inlet Connecting Pipe	ASTM A312 Type 321
Return Bend Assembly	Flange : ASTM SA182 F11 Cl.2 Elbow: ASTM A217 Gr. WC6 Pipe: ASTM SA335 P11

Codes of Construction: Equipment was designed to the respective Codes as follows:

USX-OMEGA Exchangers	
USX OMEGA Exchangers	ASME Section VIII, Division 1
USX Gas Inlet Nozzles	ASME B31.3
USX Return Bend/Gas Outlet Section	ASME B31.3
Shell/Channel/Tubes/Tube Sheets	ASME Section VIII, Division 1, TEMA Class R and API-660 Applies

Level 3 Engineering Analysis (Phase 1)

The FFS assessment followed API 579-1/ASME FFS-1, Part 10 methodology (Assessment of Components Operating in the Creep Range), starting with a Level 3 Assessment. The engineering analysis performed by KHE was based on the following parameters:

- Equipment design review
- Operating Data Analysis
  - Upset Trend Analysis
  - Identification of Key Operating Points
- Process Analysis
  - Material Properties for Process Composition
  - Energy Balance (Duty Reduction and Steam Production)
  - Thermosyphon
- Heat Transfer Analysis
  - Axisymmetric Model – Vapor Lock

- o 3D Model – Gas Side Only
- o 3D Waterside Only
- o 1st Leg Axisymmetric Inner/Outer ΔT

All of the above considerations were incorporated into an elastic-plastic finite element analysis. The results of the FEA revealed that the USX-Omega's exchangers did not satisfy the requirements of API 579/ASME FFS-1. As mentioned earlier, localized plastic strain at the inner pipe-to-header box weld area is above the threshold recommended by API 579 Part 10 of 2.5% (the strain limit at welds). The resultant stress plot is provided in **Figure 2**. Given these results, non-destructive testing and destructive analysis was pursued to determine suitability for continued service for the USX heat exchangers.

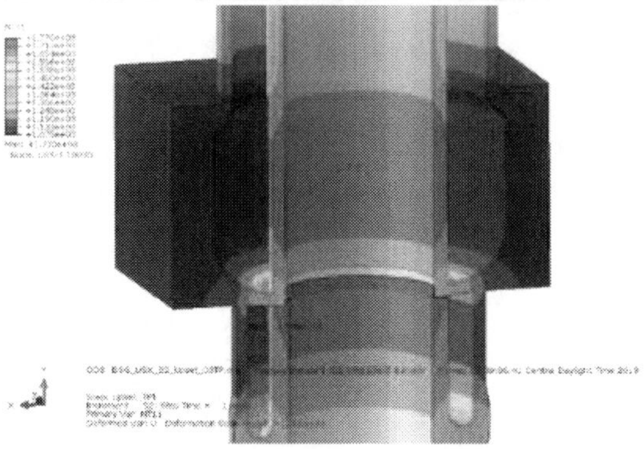

Courtesy of Knighthawk Engineering, Inc.

Figure 2 – FEA Stress Plot for Inner Pipe to Header Box Weld Area

Additionally, metallurgical analysis performed by KHE revealed that several USX-Omega return bends and primary leg outer pipes experienced temperatures outside the typical operating or design conditions. Field replication revealed that a number of the primary legs of the USX-Omega's reached temperatures outside the corresponding water saturation temperature of 532°F. FEA models suggest that several USX-Omega's secondary leg outer pipes also, reached temperatures outside the typical operating and design conditions.

As a result of KHE's Level 3 FFS analysis, non-destructive metallurgical assessment and destructive testing was pursued to qualify the equipment/components for continued service.

Acuren/MME was contracted to perform additional non-destructive testing of all USX-Omega's, as well as destructive testing of two (2) USX-Omega exchangers and its associated components (e.g. inner and outer primary and secondary piping, turbo-chambers, etc.). The metallurgical assessment and destructive testing included the following work scope:

- Field Metallurgical Assessments:
  - *In Situ* metallurgical examination with replica documentation
  - Hardness Testing
  - Dimensional Checks
- Destructive Metallurgical Analysis and Mechanical Testing

- Metallurgical cross-sections
- Bulk hardness testing
- Micro-hardness testing
- Tensile Testing
- Creep Rate Testing

Additionally, Acuren Inspection's Advanced Services Group performed ultrasonic non-destructive testing of the USX-Omega inner tubes to detect the presence of cracking due to excessive strain induced by the temperature excursion.

Non-Destructive Examination (NDE) of USX-Omega Exchangers (Phase II)

Acuren/MME performed on-site metallurgical condition assessment of the USX-Omega exchangers, as well as the coke drums and transfer line per the Inspection Test Plan (ITP) which outlined the inspection locations and the inspection work scope at each location. The on-site portion of the ITP included the following NDE techniques:

- Visual Inspections
- *In Situ* metallographic examination with replica Documentation
- Advanced Ultrasonic Testing
- Field Hardness Testing
- Dimensional Measurements

Inspection work scope included specific testing method, primary and secondary acceptance criteria at specific locations of the USX-Omega's.

Visual Inspections of the USX-Omega exchangers (after removal of insulation and cladding) were performed to identify any apparent damage (e.g. bulging, cracking, etc.) that may have occurred due to exposure to higher than expected temperatures. **Figure 3** shows examples where visual inspections were performed on the USX-Omega exchangers based upon the ITP.

Figure 3 – Visual Inspections of USX-Omega Exchangers

In situ metallographic examination at locations specified by the ITP was performed to identify evidence of thermal alteration and/or incipient creep cavitation, **Figure 4.** Additionally, if a butt weld was present, metallographic replication was performed across the weld and incorporated the upstream base metal, the weld filler, and the downstream base metal.

Intermediate Weld of Primary Leg

Secondary Leg Lower Turbo-Chamber

Primary Leg Lower Turbo-Chamber

Primary Leg Upper Turbo-Chamber

Figure 4 – Location of *In situ* Metallographic Examinations of USX-Omega Exchangers

In situ metallographic examination revealed typical base metal microstructures characteristic of carbon steel with exception to a single Secondary Inlet Turbo Chamber where some carbide decomposition was evident, **Figure 5**.

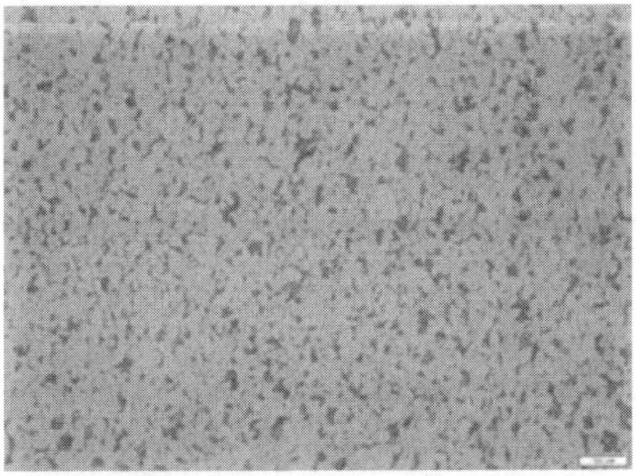

Typical Microstructure Observed on Outer Pipe

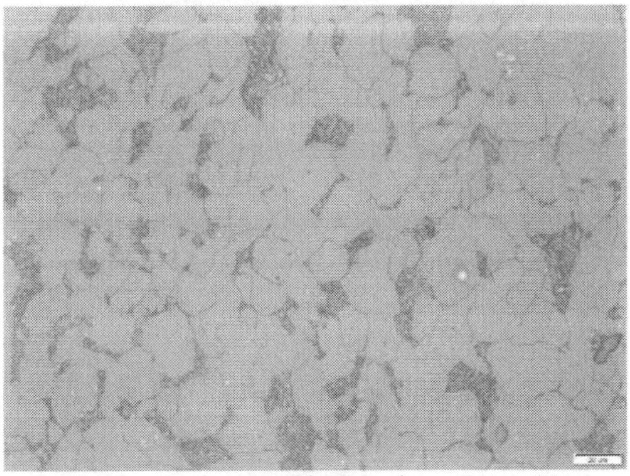

Typical Microstructure Observed on Outer Pipe

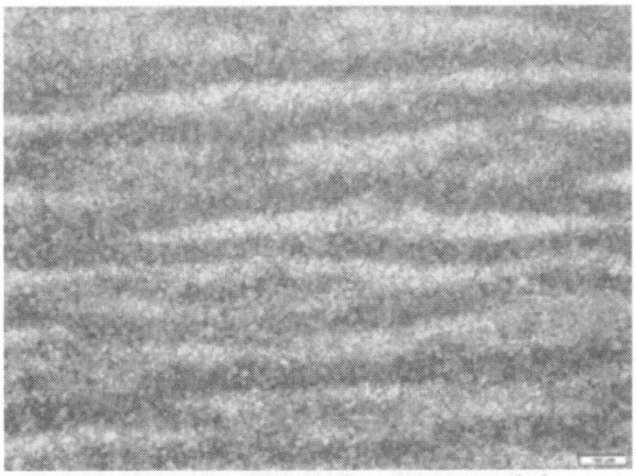

Microstructure observed at Location 8.0 on Quench Heat Exchanger A1

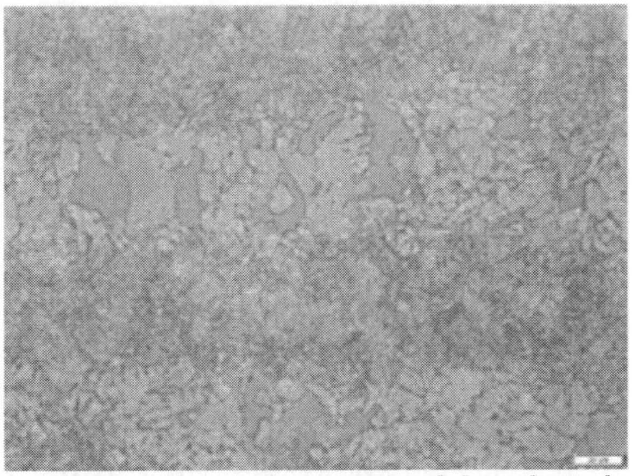

Microstructure observed at Location 8.0 on Quench Heat Exchanger A1

Figure 5 – Photographs show exemplar replica microstructures obtained from the USX-Omega Exchangers

Advanced Ultrasonic Testing of the inner tubes of both the primary and secondary legs of the USX-Omega's was performed to detect cracking due to excessive thermal strain or the possible thermal fatigue damage due to prior unit cycling. A Multi-Phased Ratio Technique (MERT) examination of the USX-Omega exchangers (Primary & Secondary legs) was performed of the accessible areas using both longitudinal and two directional shear waves. The examination was performed by line scanning the full length of each tube manually on 4 quadrants at 0°, 90°, 180°, and 270° utilizing an unrectified A-Scan display and a free running B-Scan display. This inspection was focused on the detection of degradation to the USX Omega inner tube (SA335 P1). Additionally, MERT was performed on the USX-Omega exchangers for the purpose of examining the accessible areas at the top and bottom areas of the tube and turbo-chamber header boxes. This inspection was focused on the detection of degradation to the USX-Omega inner tube (SA335 P1). **Figure 6** shows Automated Ultrasonic Testing (AUT) of the inner tube and turbo-chamber header of a USX-Omega exchanger.

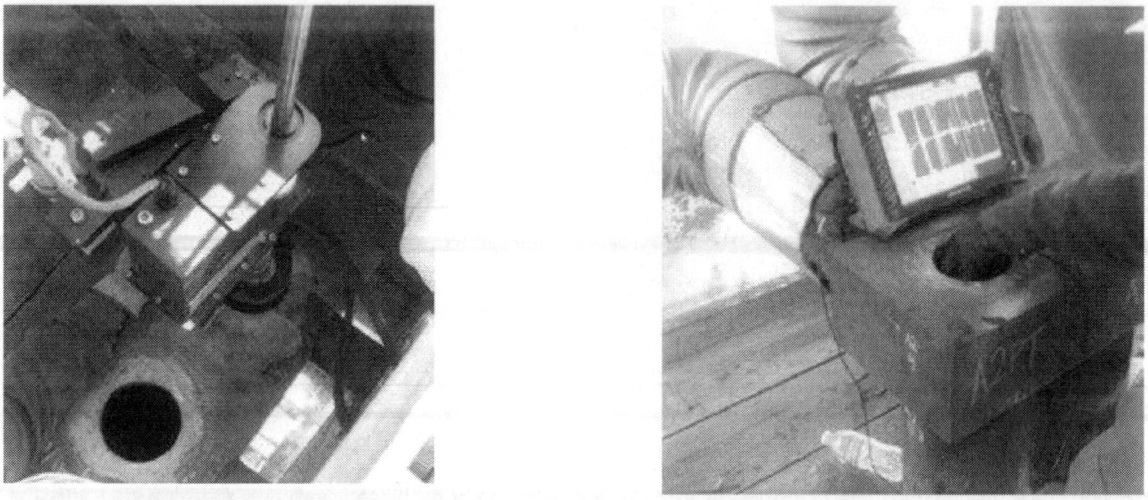

AUT of Inner Tube

AUT of Turbo-Chamber Header

Figure 6 – AUT Inspection of Inner Tubes and Turbo-Chamber Header of USX-Omega Exchangers

Hardness Testing was performed to identify softening due to the process event and to verify adequate (estimated) tensile strength of the piping, weldments, and turbo-chamber headers. Hardness Testing was performed at various locations as identified within the ITP. **Figure 7** shows a small sample of the various locations where hardness testing on the various components of the USX-Omega's were performed. The hardness lower limit was established based on the equivalent Brinell hardness per ASTM A370 conversion tables that correspond to the specified minimum tensile strength of each material used to fabricate the USX-Omega Exchangers.

Figure 7 – Typical Hardness Testing Locations of USX-Omega Exchangers

Dimensional Measurements including outer diameter measurements via PI Tape and calipers, and ultrasonic thickness measurements were performed to detect growth due to creep strain or excessive thinning was performed at various locations as identified on the ITP. Dimensional measurements did not indicate any significant pipe swell due to creep strain. Nor was excessive thinning detected as all measurements were within the appropriate mill tolerance.

Destructive Examination of USX-Omega Exchangers (Phase III)

Destructive examination of three (3) USX-Omega exchangers and its associated components (e.g. inner and outer primary and secondary piping, turbo-chambers, etc.) were performed by Acuren/MME, see **Figure 8**. The examination protocol included the following tests:

- Tensile Testing (room temperature)
- Metallographic Examination
- Creep Rupture Testing

Figure 8 – Photograph Shows Typical USX-Omega Components Received for Examination

Components were sectioned, as shown in **Figure 9** for metallographic examination, wall thickness measurements, hardness measurements, and tensile and creep testing. Samples on the left shows the outer pipe of the Omega-USX, while the photo on the right shows the inner pipe.

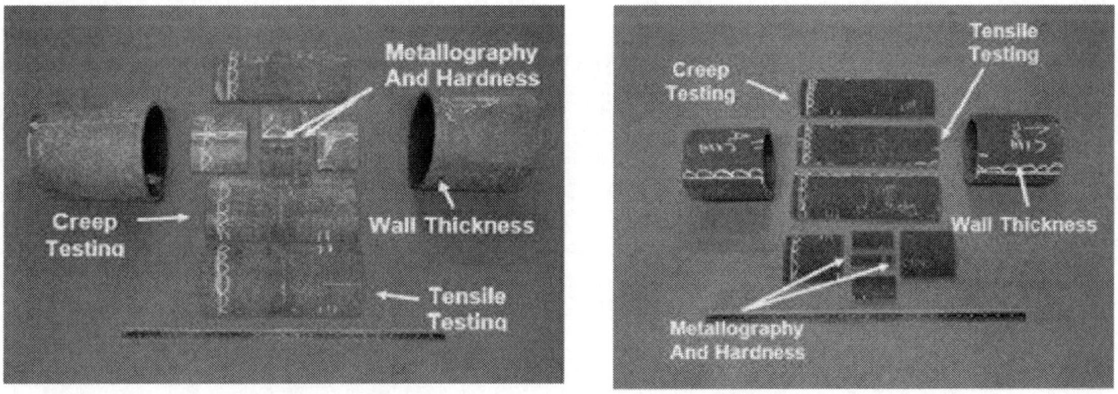

Figure 9 – Photograph Shows Components after Sectioning for Metallurgical and Mechanical Testing

Room Temperature Tensile Testing

One tensile test sample was cut from each of the USX-Omega exchanger Primary and Secondary Inner and Outer pipes, one from each dissimilar girth weld coupon, and one from the USX-Omega turbo-chambers (headers). One, longitudinal, full wall thickness, reduced section tensile (RST) specimen was removed from each section. Cross-weld tensile specimens were also prepared from the sections that contained dissimilar girth welds. The numerous samples were prepared and tested according to ASTM E8, Test Methods for Tension Testing for Metallic Materials and ASTM A370, Mechanical Testing of Steel Products. Doing so yielded typical stress-strain curves for ductile materials, as shown in **Figure 10**.

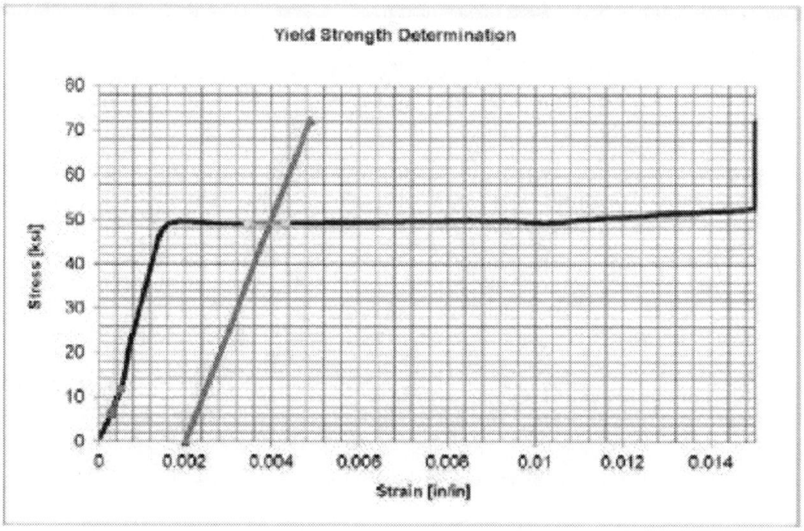

Figure 10 – Typical Stress-Strain Curve

The tensile strength of all samples was above 60 ksi. The outer piping, fabricated from ASTM A106, Grade B carbon steel, has a minimum tensile strength requirement of 60 ksi. The inner piping made of ASTM A335; Grade P1 steel has a minimum tensile strength requirement of 55 ksi.

Metallographic Examination

Transverse and longitudinal cross sections were removed from each pipe sample and were prepared for metallographic examination. The prepared cross sections were examined using a metallurgical microscope to evaluate the bulk microstructure, external and internal surface conditions, and any anomalies. **Figure 11** shows cross sections of samples after preparation.

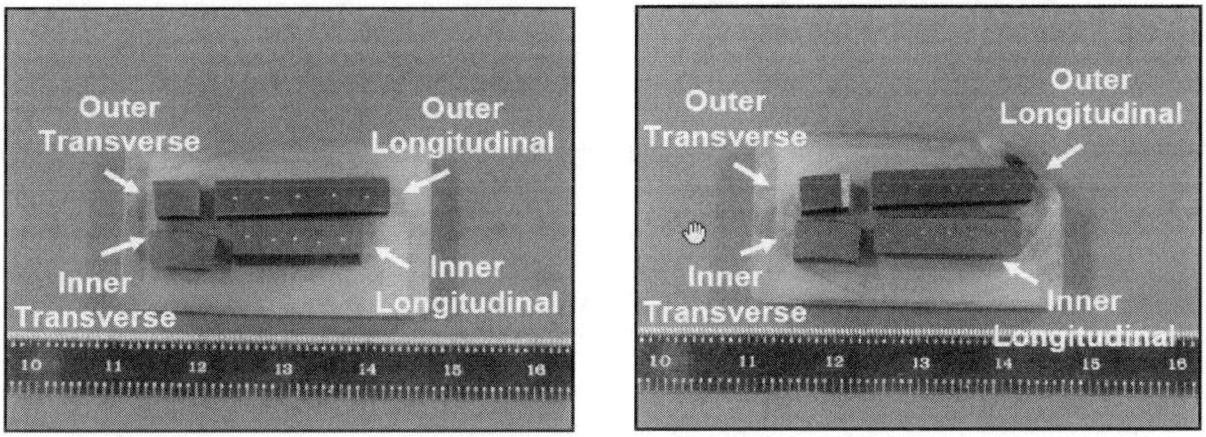

Figure 11 – Photographs of Cross Sections of Samples after Preparation

The bulk microstructures of the samples consisted mostly of ferrite and lamellar pearlite. There was no evidence of thermal degradation. All microstructures were typical of carbon steel or low alloy steel material **Figure 12** and **Figure 13**.

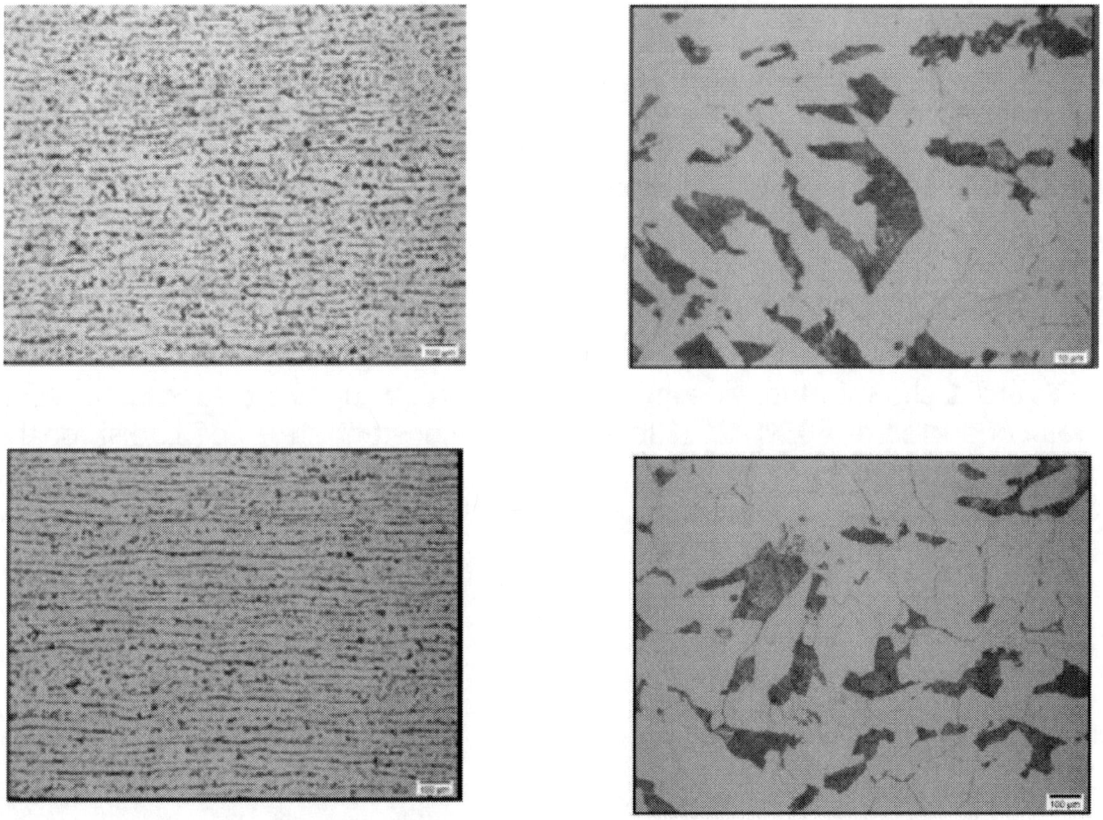

Figure 12 – Photomicrographs show the Outer Pipe Microstructure in the Transverse (upper) and Longitudinal (lower) planes

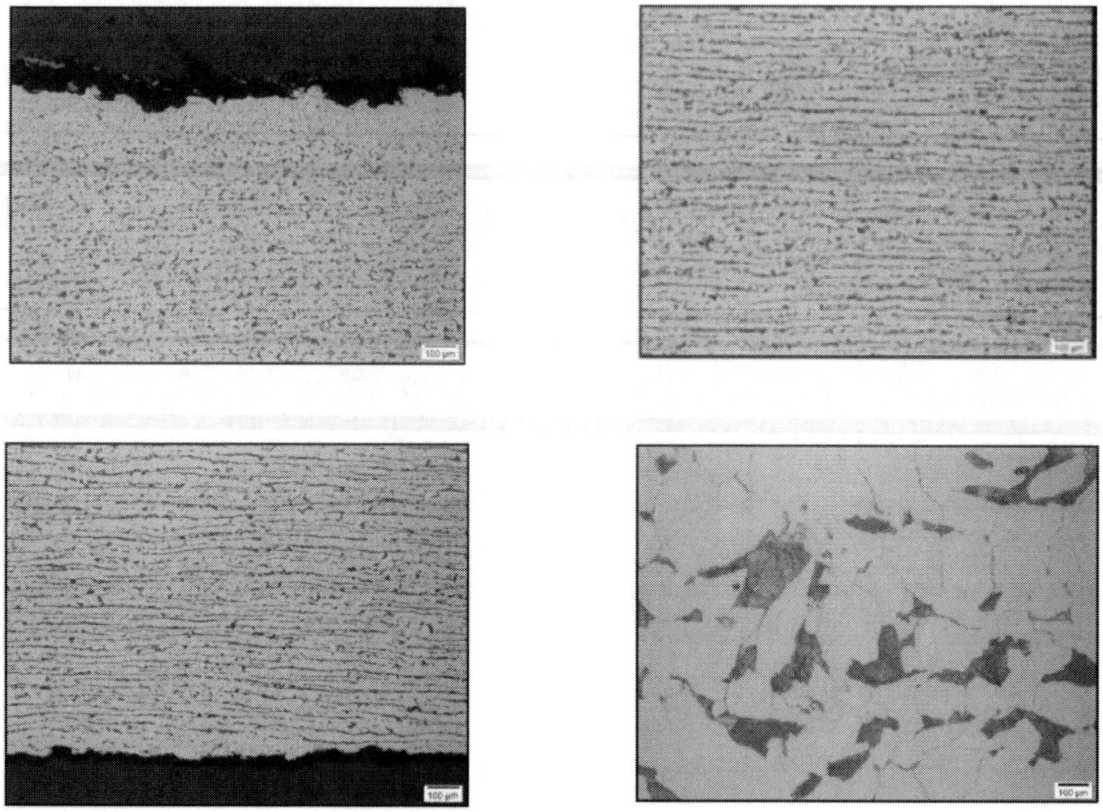

Figure 13 – Photomicrographs: Outer Pipe - Longitudinal

Creep Rate Testing (ASTM E139)

Creep rupture testing was conducted on the Primary and Secondary Inner and Outer pipes. Also included in the creep rupture testing was four dissimilar girth weld coupons and two additional samples from the USX-Omega turbo-chambers (headers).

Table 1 shows a typical summary of the creep rupture testing data results. The test was conducted at 805ºF for 100 hours at a constant stress of 12 ksi for the outer piping, weld, and header coupons, and a stress of 19 ksi for the inner piping specimens at 905ºF. Note that tests were stopped at 100 hours due to the extremely low creep strain rate accumulation indicating. This indicates that minimal to none creep life exhaustion had occurred as a result of the temperature excursion. **Figure 14** shows typical creep data of exemplar specimens.

Table 1 – Typical Creep Rupture Results

Sample ID	Temp (°F)	Stress (Ksi)	Diameter (in)	Time In Hours to Reach % Creep Strain						Test Time (Hours)	Final Strain (%)	Elongation (%)	Reduction of Area (%)	Failure Location
				0.1	0.2	0.3	0.4	0.5	0.6					
1	805	12	0.1602							100	0.078	---	---	---
2	905	19	0.1597							100	0.086	---	---	---
3	805	12	0.1604							100	0.086	---	---	---
4	905	19	0.1603							100	0.094	---	---	---
5	805	12	0.1599							100	0.094	---	---	---
6	905	19	0.1604							100	0.047	---	---	---
7	805	12	0.1604							100	0.078	---	---	---
8	905	19	0.1607							100	0.063	---	---	---

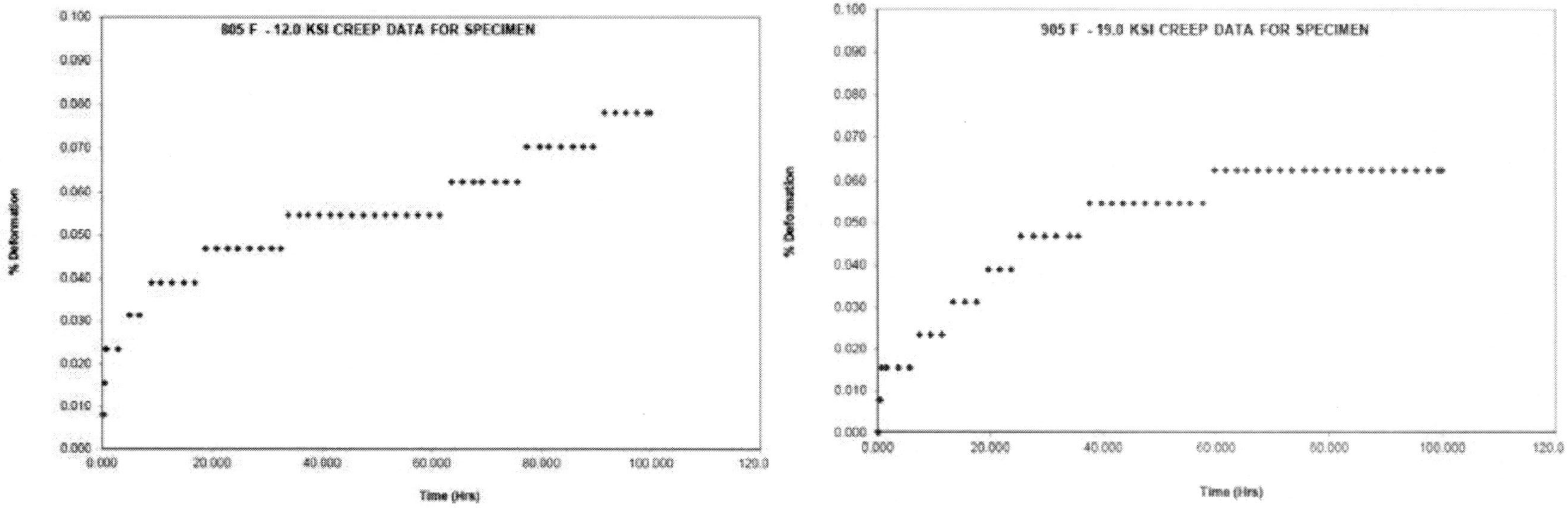

Figure 14 – Typical Creep Data of Exemplar Specimens

Conclusions

The results from the FFS Level 3 assessment indicated that the temperature excursion had indeed induced thermal strains sufficient to cause alarm regarding the reliable operation of USX-Omega heat exchangers. However, based upon destructive examinations, the USX Omega heat exchangers were deemed fit-for-continued service when operated within the equipment design parameters. The fit-for-continued service decision was further reinforced by non-destructive testing performed on the USX Omega heat exchangers.

618737 Boost Octane of Light Naphtha Isomerization Units Using Dual Dividing Wall Column Technology

Abstract:

Advancements in distillation can play a key role in minimizing costs of new and revamp projects in a refinery. Dividing wall column technology which had limited acceptance 20 years back is now enhancing our distillation processes. This technology offers varied options when it come to number of cuts that can be taken out from the column by the use of single or multiple walls.

Dual dividing wall columns have multiple walls in a single column shell. These columns have the potential to influence the profitability of most commonly used units in the refinery. One such example of dual DWC application would be a light naphtha isomerization unit. Dual DWC Technology offers the following benefits in an isomerization process:

Boost Octane using existing columns:

DWC's can boost the octane of isomerization units by 2-4 points. The isomerization process converts low octane straight chain hydrocarbons to high octane branched chain isomers. Since the isomerization reactions are in equilibrium, the product octane is defined by the number of separation units in the process.

A typical light naphtha isomerization unit with a deisohexanizer (DIH) column can produce an isomerate product of RON 87-88. For further increase in RON additional deisopentanizer (DIP) and depentanizer (DP) columns are required. Dividing wall column retrofits in the existing DIH column and provides it with an additional functionality of a DIP or a DP column.

Increase octane barrel and produce C6 products:

Deisohexanizer column (DIH column) separates light and heavy isomerate along with a side cut recycle stream to the isomerization reactor. The DIH column revamp to DWC produces an additional 4th cut of C6 products (Food Grade Hexane/ Polymer grade hexane/ Isohexane / Special boiling point solvents). This revamp not only increases the profitability of Isom unit, but also increases the capacity of Isom unit by more than 20%.

619631 Gas Concentration Absorber/Stripper Benchmarking leads to Optimization and Training Opportunity

Abstract:

Across the Marathon system there are thirty Absorber/Stripper systems spread over sixteen sites. The Absorber/Stripper systems are associated with FCC, Coker and Saturate Gas Plants. The design and operation of each unit is unique but the overall objectives and problems are similar.

A benchmarking of the FCC Absorber/Strippers were conducted using simple metrics such as Liquid/Vapor ratios, operating pressures and stage counts. The more complicated principles around water management turn out to be the overlapping difference between comparable system.

Operating curves for every FCC Absorber/Stripper system were generated and compared against units with similar designs. Process modeling was used to show how changes in controllable variables could improve performance. Typically the largest deltas in performance were explained by differences in water management.

The overall benefits from the benchmarking were associated with training, optimization and a project strategy. The clear differences in performance and water management provided an opportunity to train engineers. The monitoring of water removal and design of water removal equipment were a key training opportunities with process engineers. The operating curves provided key information around maximum/minimum constraints. This could be feedback into economic evaluations and daily optimization. The long-term improvement strategy was enhanced with improve knowledge around design and constraints.

Monitoring Benzene Precursors in Naphtha Reformer Feeds with the tec5USA Raman Spectrometer System

Authors: and Amanda Muraca[1] Norman P. Lieberman[2] and Elizabeth Lieberman[2]

[1]tec5USA, Plainview NY

[2]Process Improvement Engineering, New Orleans LA

Abstract:

The purchase of benzene credits to offset benzene levels in gasoline in excess of 0.62 vol.% is an expensive method to compensate for suboptimal control of the reformer feed. To minimize the formation of benzene, Naphtha Splitters are utilized for optimal fractionation between Benzene and Heptane (i.e. C_7's). This requires qualitative analysis of Benzene in the reformer feed and C_7's in the isomerization feed. The objective is to optimize reboiler and reflux duties, which requires a quick response for efficient control.

Using Differential Pressure (DP) or Temperature (DT) as an indicator to optimize tray efficiency is inadequate if flooding is initiated on upper trays. Using Gas Chromatography does not allow for closed loop control of the reflux due to the 10 – 25-minute response time and maintenance issues. Additionally, the cost to install, operate, and service Gas Chromatographs is significant due to the requirement of sample pretreatment and consumable accessories.

Use of the tec5USA Raman Spectrometer System and Tidus XP Immersion probe to measure the concentration of hydrocarbons in the naphtha splitter products, allows the reflux rate or reboiler duty to respond in seconds, rather than many minutes. The system allows for the simultaneous measurement of several components without additional probe installations. The maintenance requirement of the sampling system is eliminated, as the probe is directly inserted into the flowing product line through existing ¾" or 1" valves. The lockout, tagout (LOTO) system, robust design and explosion-proof enclosure allows for a safe and seamless integration to the site location and DCS/SCADA system.

The tec5USA Raman Spectrometer System and Tidus XP Probe is a cost-effective and accurate method for optimizing benzene reduction in the Naphtha Reformer feed. In contrast to other analytical methods, the Raman Spectrometer System provides quantitative results for hydrocarbon feeds in real-time with limited maintenance. This analytical method significantly reduces refinery costs due to benzene credits through direct control of the reformer and isomerization feed of the Naphtha splitter. The system is currently in service in several similar applications.

I. Introduction

Originally, the 10%-20% of naphtha material in crude oil was considered a refinery waste product and discarded. As time progressed, naphtha was utilized for gasoline fuel and lead additives (i.e. tetraethyl) were added to suppress premature detonation in the engine cylinders. However, due to environmental hazards, tetraethyl lead was replaced by MTBE additives to improve anti-knock properties (i.e. octane number). Due to its high solubility in water, MTBE degraded the quality of resources, eventually idling the MTBE plants. Thus, the process of naphtha reforming was developed, which did not require any chemical additives. This process converts naphthene into aromatics such as Benzene, Toluene and Xylene, which boost the octane rating substantially.

Although Benzene products were found to boost the octane rating considerably, benzene was found to promote cancer, and was outlawed from the inclusion of motor fuel. To this day, inclusion of Benzene higher than 0.62 vol% in motor fuel results in a significant Federal financial penalty for refineries. Those who produce more than 0.62% Benzene are required to offset this practice through the purchase of Benzene Credits. As an example, within a 100,000 BTU plant producing 50,000 BTU of gasoline at 1.62% Benzene, the cost associated with Benzene credits can be up to $10,000,000/year. Therefore, to minimize the formation of Benzene in the naphtha reformer, a naphtha splitter is utilized, as shown in Figure 1.

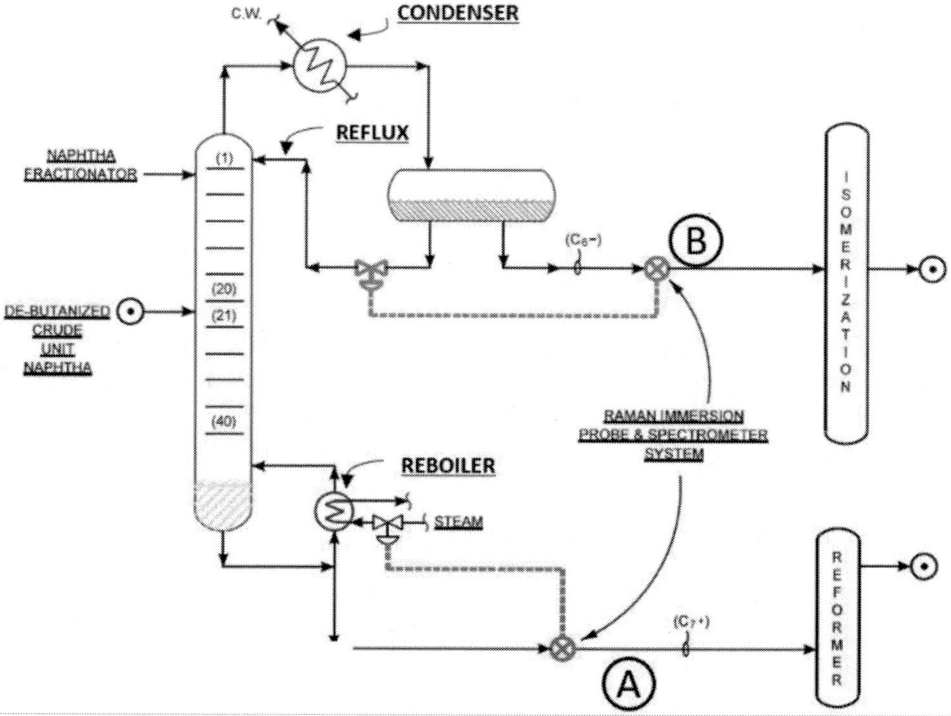

Figure 1: Naphtha Splitter process diagram. Location A and B show installation location for monitoring reformer feed and isomerization feed respectively.

The primary function of the Naphtha Splitter, within the Naphtha hydrotreating unit, is the separation of light naphtha (LN) and heavy naphtha, as described in Figure 1. As a result of the splitting process, the heavy naphtha is delivered to the Naphtha Reformer (Location A in Figure 1), where it is then converted into high-octane liquids called reformates.[1] In accordance with the Federal Regulations previously described, it is critical to minimize the concentration of benzene precursors which are present in the Reformer feed.

In addition to Naphtha Reforming, there is an alternate Refinery Process, used for smaller molecules to upgrade their octane. This process is called "isomerization" and the feed is shown by Location B in Figure 1. An early example of this process was used at the American Oil Refinery in Indiana. This unit converted Normal Butane (used to increase the vapor pressure of gasoline) into Isobutane (used as feed for an Alkylation Unit). Unlike the reformer, the isomerization unit does not produce net hydrogen for export.

The negative aspect of the isomerization unit is that it primarily converts molecules with more than six carbon atoms to ethane, propane, and butane, rather than to high octane gasoline. Additionally, the octane number boost for hexane is not as significant (i.e. molecules with six carbon atoms). However, the isomerization unit does not create benzene from the naphthene rings and therefore, benzene penalties do not introduce an issue for the isomerization feed. Overall, the Naphtha Splitter serves two primary functions:

1. Minimize benzene precursors in the Reformer feed (i.e. C_6 Naphthenes)
2. Minimize molecules having seven or more (i.e. heptanes) carbon atoms in the Isomerization Unit feed.

The addition of top reflux to a distillation tower creates more liquid traffic onto the fractionation tray. The liquid cascades down through the trays, into the bottom of the fractionator, and into the reboiler. As a result, the reboiler TI sensor will detect the declining temperature and automatically increase the heating steam flow to the reboiler, re-vaporizing the reflux. Newly generated vapors then bubble upward through the fractionation trays and into the tower's overhead condenser, where the vapors are restored to the liquid phase – i.e. reflux. The net effect is to improve the fractionation between benzene precursors and heptane (i.e. C_6's and C_7's) by forcing the trays to work harder.

However, as the vapor flow through the trays increases, the enhanced vapor velocity tends to promote entrainment (droplets rising from the tray below to the tray above) called "Jet Flooding". At some point, this entrainment will overwhelm the beneficial effects of reflux. When this first happens, it is referred to as the "Incipient Flood Point". At this operational point in a fractionator, increasing reflux (or reboiler steam) in fact worsens fractionation rather than improving it.

Traditional methods for monitoring key parameters within the refinery, and more specifically within the Naphtha Splitter, include differential pressure, differential temperature, laboratory analysis and online Gas Chromatography (GC). These techniques introduce drawbacks such as inaccuracy, significant lag-time, and high maintenance. In contrast to these methods, Raman Spectroscopy, with the tec5USA Raman Analyzer, is a cost-effective, immediate, and accurate method for optimizing the performance of the Naphtha Splitter and improving the fractionation of Benzene and Heptane (C_7's), by providing immediate and accurate chemical analysis at the point of insertion.

Raman spectroscopy is a non-destructive technique that utilizes a laser source to interact with and excite molecules of a sample (Figure 2). The interaction with these molecules results in a frequency (energy) shift of the laser photons which are reported in wavenumbers (cm^{-1}). This frequency shift or "Raman Shift" is dependent on how the incident laser light interacts with the molecules and provides information regarding the vibrational modes of the species. For each molecular species, analysis of these different frequencies can be used as a "fingerprint", where each molecule has its own unique spectrum. The spectral result is used to qualitatively identify the composition of a sample material based on this unique fingerprint. Furthermore, a quantitative value for these components is derived from the resulting signal intensity and chemometric calibration methods.[2] Raman spectroscopy has been utilized for quantitative analysis of various parameters, specifically within refinery and petrochemical applications.[3,4] The tec5USA Raman Analyzer System is a cost-effective and accurate method for optimizing benzene reduction and improving the performance of the Naphtha Splitter.

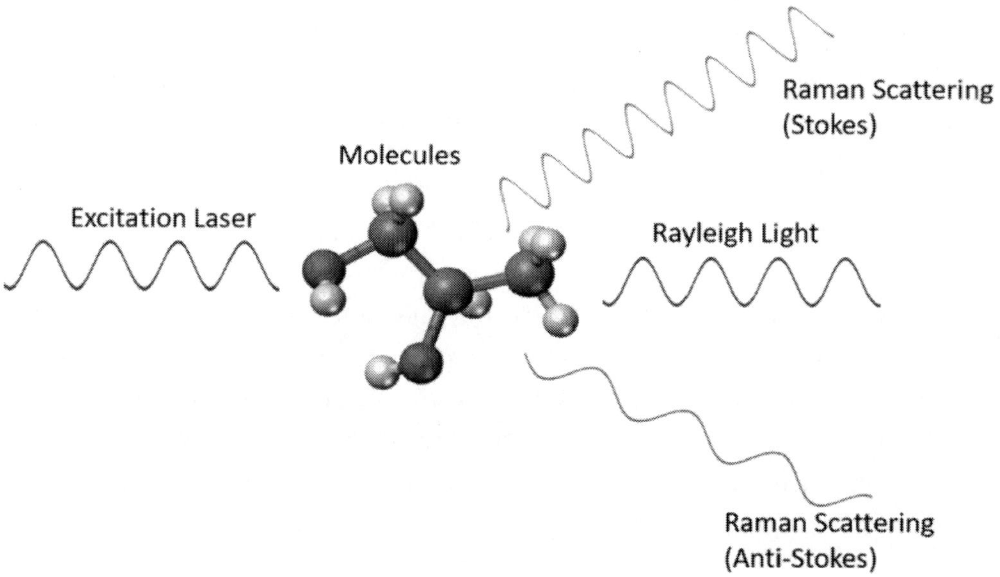

Figure 2: Fundamental principle of Raman spectroscopy, where the excitation laser interactions with molecules and causes Stokes and Anti-Stokes frequency shifts.

II. Optimizing Reflux Rates in Fractionators

The objective of the on-stream optimization of reflux rates and reboiler duties, is one of the main objectives of Chemical Engineering and certainly of Refinery Process Engineering. Likely in the experience of 90% of process engineers, optimization of the fractionator reflux and reboiler duties would be their primary technical objective when working with fractionators. Methods typically employed:

1. Differential Pressure – In theory, the optimum efficiency of a fractionation tray, set on 24" tray spacing, is 4" – 6" of liquid DP. However, this relationship may be rendered useless, because of formation of foam on the tray decks. Also, we usually measure a single differential pressure across a tower with 30 – 50 trays. The flooding may start at some intermediate tray, and then the overall DP does not mean much.
2. Differential Temperature – In theory, the greater the DT between the top and bottom trays, the better the fractionation. However, once a section of a tower becomes involved in "Fully Developed Flood", the liquid being pushed up the column, can cause a decline in the top temperature, even though the trays are flooding.
3. Lab Samples – In most fractionators, the operators rely on laboratory samples. From the time the operator on the process control panel decides to get a check sample, one or two hours will pass, until the lab result is available for the operator to adjust the reflux rate, or the steam flow to the reboiler. By that time, the fractionator's feed composition may have changed, or tower pressure, or feed temperature, may have altered.
4. On-Line Gas Chromatograph Control – We have used this technique with good results at several locations:
   a. Adjusting steam flow to a Fluid Cracking Unit De-ethanizer steam reboiler.
   b. Adjusting the reflux on a Saturated Propane – Butane splitter, to produce HD5 – propane, and Alky Unit feed.

However, the complications associated with use of the Gas Chromatographic controls include:

1. Cost – Typically $300k - $500k installed
2. Complexity of the installation.
3. Instrument maintenance – including carrier gas and column consumables
4. Analysis lag time of 12 – 25 minutes
5. Reliability of the analyzer sample system.

III. tec5USA Raman Analyzer and Tidus XP Immersion Probe

In contrast to traditional methods, including differential pressure, differential temperature, laboratory analysis and on-line Gas Chromatography (GC), the tec5USA Raman Spectrometer system serves as an improved methodology for optimizing reflux rates and reboiler duties. The signal which is generated by Raman probe is used to report direct results to the automation system for closed-loop control via seamless connection to DCS and/or SCADA systems. Use of the tec5USA Raman Analyzer and Immersion probe provides quantitative information of the hydrocarbons within the naphtha splitter products.

The tec5USA Raman Analyzer solution consists of a Field Raman Spectrometer System (Figure 3a) and the Tidus XP Raman Immersion Probe (Figure 3b). The lockout tagout (LOTO) system, robust design and explosion-proof enclosure allows for safe integration into the harsh environments of the refinery. The tec5USA Raman Analyzer utilizes a temperature stabilized 785 nm laser with tunable output from 35 – 500 mW. Allowing for safe laser operation below any ignition energies of any chemical species present in the environment. The system contains a highly sensitive CCD array used in conjunction with VPH transmission gratins, allowing for straylight reduction, highly efficient dispersion, and a wide spectral range of 300 – 3100 cm^{-1}. The fiber-bundle cross section converter optimizes light throughput by forming a linear arrangement of fibers to create the spectrometer entrance slit. As indicated in Figure 1, this analytical tool may be positioned at location "A", specifically to identify the benzene content (vol%) in heavy Naphtha before sending the feed to the reformer. The primary goal of this analysis is to formulate a suitable feed for the reformer by minimizing benzene precursors in the heavy Naphtha feed (i.e. C_6 Naphthene). As a result, the quantitative concentration of benzene is reported directly to the reboiler steam flow control valve, enabling closed-loop control of benzene in the reformer

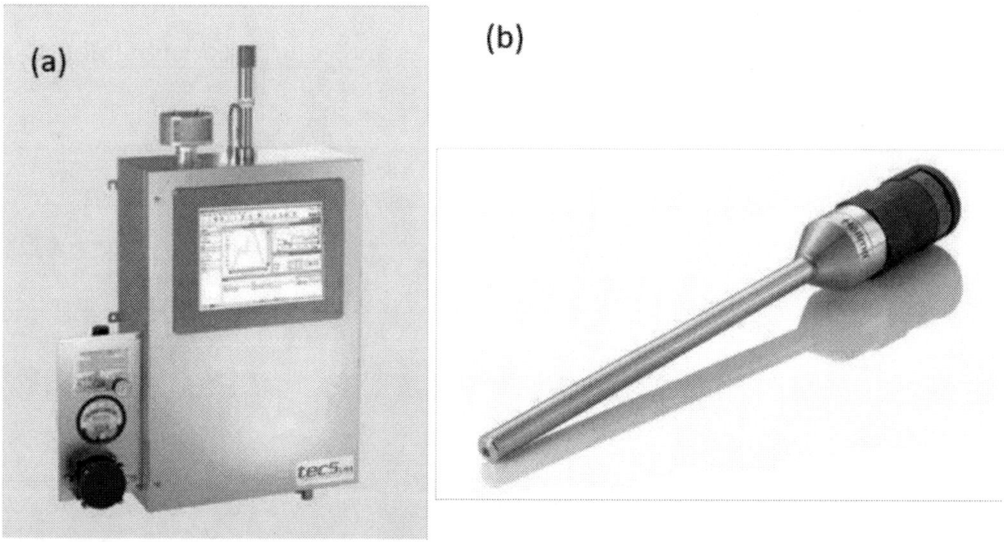

Figure 3: tec5USA Online Raman Analyzer system (a) Raman Analyzer Field Unit, housing the light source, spectrometer, and electronics (b) Tidus XP Raman Immersion Probe for direct insertion into the process stream.

feed. Location "B" can also be utilized in conjunction with location "A" for process optimization. Shown in Figure 1, analysis at this location allows for the quantitative measurement of molecules with 7 carbon atoms or greater in the Isomerization unit feed. Concentration levels of C_7's and C_{7+}'s are reported directly to the Reflux by the tec5USA Raman Analyzer, where corrective actions may be taken to minimize C_7's and C_{7+}'s within the feed.

As previously discussed, the fractionator can often operate above the "Incipient Flood Point", causing "Jet flooding" and degrading the performance of the fractionator. In this occurrence, additional reflux and/or reboiler steam can worsen fractionation. The onset of this "Incipient Flood Point" can be immediately identified with the tec5USA Raman Analyzer. In conjunction with closed-loop control, immediate action to halt the Reflux or Reboiler may be taken. Alternatively, adjustments to reduce the reflux rate and/or reboiler duty can be automated based on the feedback from the tec5USA Raman Analyzer. In addition to providing analytical results which are precise, accurate and report to existing control valves, the tec5USA Raman Analyzer also provides the following advantages over traditional methods, such as Gas Chromatography.

COST – At a price-point of <$100k, the tec5USA Raman Analyzer is a cost-effective alternative to standard methods and eliminates the need for offline, laboratory analysis. For installations where multiple measurement locations are required, optical multiplexing serves as a cost-efficient alternative to multiple individual analyzers. With optical multiplexing, the tec5USA Raman Analyzer can measure up to 4 different locations with a single spectrometer. As shown by the Schematic in figure 4, 4 Raman Immersion Probes can be positioned at 4 different

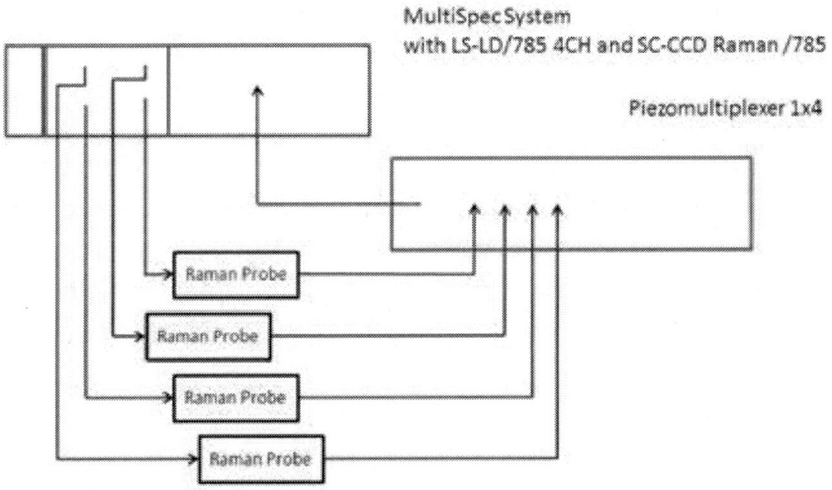

Figure 4: tec5USA Online Raman Analyzer 4-Channel Multiplexing Schematic

measurement locations (within 100 m of a central location). The signal from each probe is delivered to a single Raman analyzer with fiber optic cables. The Multiplexer utilizes a Piezo multiplexer switch (see Figure 5), where a translation stage moves between two end points. This technology allows for a fast switching time between channels of approximately 50 - 100 ms and therefore introduces no lag-time between each measurement. The translational stage also allows for nearly 100% accuracy, contributing to the overall high performance and accuracy of the tec5USA Raman Analyzer. Overall, this setup enables significant cost reduction per measurement channel.

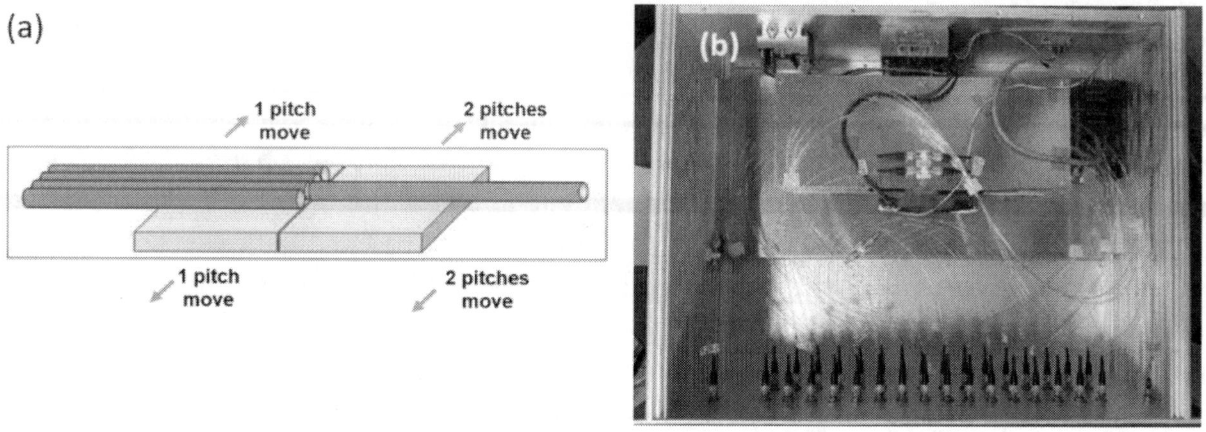

Figure 5: tec5USA Multiplexing system (a) Piezo multiplexer principle with a moving stage, allowing for nearly 100% alignment accuracy (b) internal example of Multiplexer with 32 channels.

EASE OF INSTALLATION – A major advantage of the tec5USA Raman Analyzer is the "Tidus XP" Immersion Raman probe. In contrast to traditional Gas Chromatography, where the sample is pre-processed and delivered to the system, the Raman probe (Figure 3b) is directly inserted into the process. As shown in figure 6, The 20 mm barrel diameter allows the probe to be immersed into the process flow on-stream through existing ¾" – 1" sampling ports, either within the vessel itself or through the pipeline As there are multiple options available for probe length, installation to a variety of pipe diameters, between 3" – 10", is possible. The resulting optical signal from the measurement is transported to the Raman spectrometer via fiber optic cabling, eliminating the need for a complex sampling system.

LOW MAINTENANCE – In addition to the ease of installation, the tec5USA Raman Analyzer solution requires minimal maintenance. As previously discussed, the tec5USA Raman Analyzer is based on the principle of optical spectroscopy, where the analysis is performed by quantifying laser interaction with the sample material. Due to this principle, the technique is non-destructive and requires no pre-conditioning of the sample. Therefore, the labor-intensive procedure of replacing the GC carrier gas cylinders (e.g. Helium or Neon) and replacing the GC column does not apply to the tec5USA Raman Analyzer. Furthermore, the installation and calibration of the device is handled by the tec5USA service team. The team coordinates with the end users and operators to efficiently commission the Raman Analyzer on-stream so as not to cause a disturbance to the production process. As an added service, tec5USA offers an annual service agreement ($2k) to ensure the device is operating in accordance with specifications.

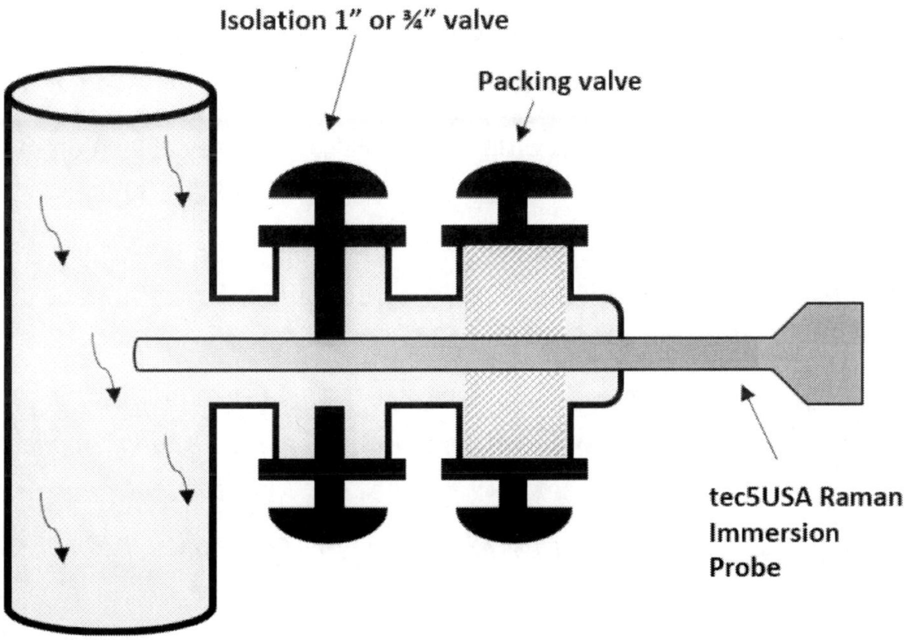

Figure 6: Sketch of the tec5USA Raman Immersion Probe installation to existing 1" or ¾" valve, allowing for probe removal.

MESUREMENT SPEED – GC lag time does not allow for true closed loop control as corrective actions are not taken immediately. In other words, by the time the result from the GC analysis is determined (~25 minutes), the sample will have likely moved to the next phase of the process. With tec5USA Raman analyzer, the quantitative results are reported within seconds. In the case of the Naphtha Splitter, these results can be transmitted directly to reboiler steam flow control valve or to the Reflux valve for adjustments and allow for a response within seconds, rather than several minutes. Moreover, the fast response time of the tec5USA Raman Analyzer can immediately detect when the "Incipient Flood Point"[5] is reached and instantly shut down or reduce reflux and reboiler steam.

IV. Conclusion

Benzene credits introduce a significant cost for refineries across the United States. Therefore, solutions such as the introduction of a Naphtha Splitter, aid to reduce benzene and optimize the reforming process. Traditional methods to improve performance of the Naphtha Splitter are often inaccurate, cost intensive or do not provide true closed loop control. The tec5USA Raman Analyzer, coupled with the Tidus XP Immersion Probe, allows for inline monitoring of the Naphtha Splitter. Additional attributes such as the low cost, ease of installation, low maintenance and real-time measurement speed, provide a true advantage over current analytical techniques, such as Gas Chromatography. Providing a response in seconds, rather than minutes, the tec5USA Raman Analyzer allows for immediate actions from the reflux or reboiler, optimizing fractionation, minimizing heptane in the isomerization feed and reducing benzene levels in the reformer feed.

V. References

1. Nevado, A., Oñate, L., Sanz, J., Alcalde, R., & Slaven, K. (2008). Temperature Optimization of a Naphtha Splitter Unit. *Advances in Electrical and Computer Engineering, 8*, 7-16.
2. Thomas, E.V., Haaland, D.M. (1990). Comparison of multivariate calibration methods for quantitative spectral analysis. *Analytical Chemistry*. 62 (10), 1091-1099.
3. Zhang, Nai & Tian, ZuoJi & Leng, YingYing & Wang, HuiTong & Song, FuQing & Meng, JianHua. (2007). Raman characteristics of hydrocarbon and hydrocarbon inclusions. *Science in China Series D: Earth Sciences. 50*, 1171-1178.
4. Buldakov, M. & Korolev, B. & Matrosov, I. & Petrov, Dmitry & Tikhomirov, A.. (2013). Raman gas analyzer for determining the composition of natural gas. *Journal of Applied Spectroscopy.* 80.
5. Lieberman, N.P. Troubleshooting Process Operaton – 5th Ed. *PennWell Publications, Tulsa.*

619649 Separation Process Essentials – A New Book and Approach to the Introductory Course

Abstract:

My unique new textbook targets undergraduate chemical engineering students being introduced for the first time to staged separation processes. As the title suggests, the content is limited to the essential skills that can be taught in a one-semester course. For each of the major separation processes (distillation, absorption/stripping, and solvent extraction) students learn to analyze (when design is specified) the process performance by solving the governing equations, and then design (when performance is specified) a process using graphical techniques.

No matter the process, the approach is identical. Each stage is described by a surprisingly limited set of equations that the students learned in the previous year: material and energy balances, phase equilibrium, and usually some process specifications. Students draw a labeled process flow diagram, perform a degree of freedom analysis by comparing the number of unknown variables to the number of appropriate equations, and solve those equations using Excel's Solver function. It is really that simple! More stages – just more equations for Excel to solve. Students then plot the compositions on a phase equilibrium graph and "discover" for themselves the McCabe-Thiele and Hunter-Nash graphical methods. Each chapter starts with a very brief introduction, but most learning occurs in the plethora of detailed examples, complete with well-formatted Excel solutions (distributed on an open web site "separationprocess.com").

In this talk/poster, the book and web site will be described in more detail. Several example problems will demonstrate the book's approach.

619655 An Investigation of Practical Implications of Complex Column Arrangements in Industrial Methanol Distillation

Abstract:

In industrial Methanol distillation, a stream consisting of a multi-component mixture of Methanol, Water and a number of trace components must be refined into AA grade Methanol, containing less than 10 wt-ppm of the middle-boiling Ethanol. Current Methanol production economic objectives also dictate that close to 100% of Methanol entering the column is recovered as product, while ensuring all hydrocarbons (primarily consisting of Methanol not being recovered as product) is recycled back to the reforming process to make syngas. In the recent past these requirements have prompted some engineers to question the "status quo" design of using a distillation column with a side draw to achieve these operational objectives, particularly with regards to energy efficiency.
The objective of this paper is to study the energy efficiency benefits that could be realized by different column configurations. To this end five practical column configurations that include a divided wall column, a recovery column arrangement, a side draw column, a cut divided wall column as well as a direct synthesis configuration were analyzed with the aid of an industrial process simulator. All configurations investigated were able to match the strict industrial product specifications and operational objectives. The energy usage of the column configurations were improved through a systematic analysis and the results revealed that all intensified columns have a superior energy usage compared to the direct synthesis configuration as expected, but with the gains limited to only 3%. An unbiased consideration of the energy efficiency results, together with other external factors including cost of capital and operability, shows that the recovery column configuration would be the most practical choice despite the presence of the divided wall column configuration which represents a greater level of intensification and integration.

619651 Procedures and Practices for the Development and Use of Physical Property Models

Abstract:

Process modeling is widely used in conceptual design, detailed engineering, plant startup, and plant operations and maintenance. In turn, process modeling is enabled by models that capture the thermodynamic properties and phase behavior of the various chemical compounds in the process, whether they are the feedstocks, products, or undesired impurities. These property models must be reliable and accurate since they determine the modeling results. In addition, property models can serve as a virtual laboratory. These procedures and practices are elucidated through two cases studies: (1) Water dew Point; and (2) Methanol Purification.

619652 Distillation Column Internal Performance Testing

Abstract:

Fractionation Research, Inc. (FRI) provides data to its membership via running a similar experiment when testing new devices or modifications to existing devices. FRI data allows members to better decide what devices to use in their existing columns and/or how to design a new column in an optimum manner. The objective of this paper is to explain to the audience how the data at FRI is obtained before it is converted into efficiency curves, performance diagrams, and correlations. An overview of the typical experimental procedure will be discussed.

617379 A NOVEL Configuration for Multicomponent Batch Distillation

Abstract:

Abstract:

While the dividing wall columns for continuous distillations has become quite popular, in this paper, we present the first detailed analysis of a dividing wall column for batch distillation. The mathematical model for a dividing wall side stripper batch distillation column is presented. Since there are many operating parameter combinations possible for this column, selecting the right operating parameters according to product specifications is presented as an optimization problem. This new configuration's performance is compared with the conventional batch rectifier and the middle vessel columns using two case studies related to the separation of a ternary mixture. The new column performance is significantly better than the rectifier or the middle vessel column. As the purity specifications increase, the percentage increase in profit due to the new configuration compared to the rectifier and the middle vessel increases.

Bibliography

Agrawal, R., Batch Distillation, US Patent 6,550,274 B1(2003)

Agrawal R. Multicomponent columns with partitions and multiple reboilers and condensers. Ind Eng Chem Res., 40(20):4258–4266 (2001)

Agrawal, R. and Z. T. Fidkowski, New Thermally Coupled Schemes for Ternary Distillation, AIChE Journal, 45, 485 – 496 (1999)

Chen Z. and R. Agrawal, Classification and Comparison of Dividing Walls for

Distillation Columns, Processes 8 (2020) 699; doi:10.3390/pr8060699.

Diwekar U., Batch Distillation: Simulation, Optimal Design and Control, 2nd edition, CRC Press, 2012.

Diwekar U. , Introduction to Applied Optimization, 3rd edition, Springer, 2020.

Dejanovic' I., L. Matijaševic', Zˇ. Olujic', Dividing wall column – A breakthrough

towards sustainable distilling, Chem. Eng. Process.: Process Intensification 49

(2010) 559.

Devidyan A. G., V. N. Kiva, G. A. Meski, and M. Morari, Batch distillation in a column with a middle vessel, Chemical Engineering Science,

49 (1994) 3033.

Fidkowski, Z. T., and L. Krolikowski, Minimum Energy Requirements of Thermally Coupled

Distillation Systems, AIChE J., 33, 654 (1987)

Harmsen J., Process intensification in the petrochemicals industry: drivers and hurdles for commercial implementation, Chem. Eng. Process.: Process Intensification 49 (2010) 70.

Hasebe S., M. Noda, I. Hashimoto, Optimal operation policy for multi-effect batch distillation system, Comp. Chem. Eng., 21 (1997) s1221.

Kaibel G., Distillation columns with longitudinal subdivision, Chem. Ing. Tech. 59 (1987) 533.

Kaibel G., Distillation columns with vertical partitions. Chem Eng. Tech.,10(1):92–98 (1987)

Li X., T Zhao, Y. Wang, Z. Zhu, process systems engineering operational design and improvement of conventional batch distillation and middle-vessel batch distillation, Braz. J. Chem. Eng., 35 (2018) http://dx.doi.org/10.1590/0104-6632.20180352s20160522.

Lorenz H., D. Staak, T. Grutzner, and J. Repke, Divided wall columns: usefulness and challenges, conference paper in chemical engineering transactions, 69 (2018).

MultibatchDS, http://www.vri-custom.org/MultiBatchDS_1532944845.pdf, accessed August 27, 2020.

Parkinson C., Dividing-wall columns find greater appeal, Chem. Eng. Prog. 103 (2007) 8.

Petlyuk F. B., V.M. Platonov, D.M. Slavinskii, Thermodynamically optimal method for separating multicomponent mixtures, Int. Chem. Eng. 5 (1965) 555.

Ramapriya, G., Tawarmalani, M., Agrawal, R., A Systematic Method to Synthesize All Dividing Wall Columns for n-Component Separation—Parts I , 64(2), 649-659 (2018), Part II, 64(2), 660 – 672 (2018)

Schultz M.A., D.G. Stewart, J.M. Harris, S.P. Rosenblum, M.S. Shakur, D.E. O'Brien, Reduce costs with dividing-wall columns, Chem. Eng. Prog. 98 (2002) 64.

Skogestad S., B. Wittgens, R. Litto, and E. Sorensen, Multivessel Batch Distillation, AIChE J. 43 (1997) 971.

Wright R.O., Fractionation Apparatus. US Patent 2,471,134 (1949)

Yildirim O., Anton A. Kiss, and Eugeny Y. Kenig, Dividing wall columns in chemical process industry: A review on current activities, Separation and Purification Technology 80 (2011) 403.

618518 Improved Estimation of Constant Relative Volatilities to Identify Energy-Efficient Heat-Integrated Distillation Configurations

Abstract:

When performing a multicomponent separation, the search space of distillation configurations is vast [1]. Although well-established commercial process simulators can be used to analyze each configuration, this is a tedious and time-consuming task. Hence, industrial practitioners often rely on heuristics and past experience when choosing a configuration. An alternative methodology, which is faster than performing detailed simulations and can yield more attractive configurations than heuristics, is to utilize shortcut methods [2]. Shortcut methods can be employed within an optimization formulation in order to systematically comb through the numerous configurations and quickly identify a handful of attractive candidates.

One of the challenges during the optimization of a distillation configuration is the complex and inter-connected vapor-liquid equilibrium (VLE) equations. Shortcut methods (such as the classical Underwood and Fenske methods) often assume constant relative volatilities to overcome this computational challenge. However, relative volatilities can vary widely for a multicomponent mixture. Building upon prior works [3,4], we propose a three-step parameter estimation approach to determine the best set of values for this assumption [5]. The key insight is to treat relative volatility as a mathematical parameter, whose value is to be determined by fitting so that the surrogate VLE is as close as possible to the true VLE. We demonstrate that this approach generates relative volatilities that can effectively capture the true VLE for a four-component paraffins mixture.

To separate this mixture, we identify energy-efficient distillation configurations which utilize heat integration, by using the generated relative volatilities in an optimization formulation that we developed. During the modelling of heat integration, feasibility of heat transfer must be checked, and is traditionally performed by comparing temperatures. However, temperatures are costly to calculate during optimization. Taking advantage of the constant values of relative volatilities in our formulation, we develop shortcut criteria to check for feasibility of heat transfer in a manner that is computationally cheaper than the use of temperatures.

1. Shah, V. H., & Agrawal, R. (2010). A Matrix Method for Multicomponent Distillation Sequences. *AIChE Journal*, 56(7), 1759–1775.
2. Madenoor Ramapriya, G., Selvarajah, A., Jimenez Cucaita, L. E., Huff, J., Tawarmalani, M., & Agrawal, R. (2018). Short-Cut Methods versus Rigorous Methods for Performance-Evaluation of Distillation Configurations. *Industrial & Engineering Chemistry Research*, 57(22), 7726–7731.
3. Anderson, N. J., & Doherty, M. F. (1984). An Approximate Model for Binary Azeotropic Distillation Design. Chemical Engineering Science, 39(1), 11-19.
4. Tumbalam Gooty, R., Mathew, T. J., Tawarmalani, M., & Agrawal, R. An MINLP Formulation for Identification of Thermodynamically-efficient Multicomponent Distillation Configurations. *In Preparation.*
5. Mathew, T. J., Tumbalam Gooty, R., Tawarmalani, M., & Agrawal, R. Quickly Assess Distillation Columns. *Accepted.*

Good Distributor Design for High-Velocity Feed Debottlenecks a Crude Preflash Tower

By

Henry Z. Kister, Fluor, Aliso Viejo, CA, USA,

Betzalel Blum, Bazan Group, Haifa, Israel,

and

Raymond Tsang, Koch-Glitsch, UK

Presented at the Distillation Symposium, AIChE Spring Meeting, Virtual, April 18-22, 2021

Summary

Bazan's Crude Unit No. 3 (designed by others) was bottlenecked by the preflash tower. The tower experienced frequent episodes, especially with light crudes (high naphtha percent), in which the naphtha stream leaving this tower turned dark, accompanied by a high tower ΔP. The episodes, most likely, originated from crude entrainment. These episodes restricted the throughput of the entire crude unit.

From the lowest downcomer, a 4" pipe drains the liquid into the top shed deck. At two recent outages, the 4" pipe was found broken and falling down, with a hole in the seal pan from which it originated. As the pipe was located right in front of a high-velocity feed entry, it was likely that feed impingement at high velocities generated vibration and caused the 4" pipe to break. It was believed that due to the pipe breakage, the high feed vapor velocities picked up the descending liquid or rose up the bottom downcomer, causing the naphtha product to become dark. The initial solution sought was reorientation of the 4" pipe away from the feed.

Suspecting that there may be other issues, Bazan requested Fluor to assist with the problem analysis. Koch-Glitsch were later added to the team to consult on hardware changes. Our analysis determined that the breakage of 4" pipe was insufficient to explain the entrainment and dark naphtha. We determined that the root cause was excessive feed velocities, with feed entry too close to the bottom tray.

To solve the problem, it was necessary to break the incoming momentum and to divert the feed to a lower point in the tower. This was achieved by a unique feed entry design for the very high-velocity feed, by removing two segments of the top shed deck to permit feed entry at a lower point, and by adding a specially-designed impingement baffle to break the incoming momentum.

When the tower returned to service, the dark naphtha episodes completely disappeared. The naphtha product rate could be raised significantly, by 20% and above, without any adverse effects on operation. No limit has been reached. The entrainment problem has been eliminated.

Our paper details the problem, analysis, and solution.

Process and Problem Description

The preflash tower (Figure 1) is designed to receive crude feed at 160°C, which contains 4.8% vapor by weight. The tower operates at a pressure of 1.6 barg. Overhead vapor from the tower is condensed, some returned to the tower as reflux, the rest is the naphtha overhead product that is sent to the main stabilizer tower. Bottom product is preflashed crude that goes to heat exchangers, the crude heater, and finally to the atmospheric tower. The 2.1 m diameter preflash tower contained 12 single-pass rectangular moving valve trays above the crude feed inlet and 6 shed decks with 34% open windows beneath the feed inlet. There is a connection that permits stripping steam to be added below the stripping section, but this steam has not been used.

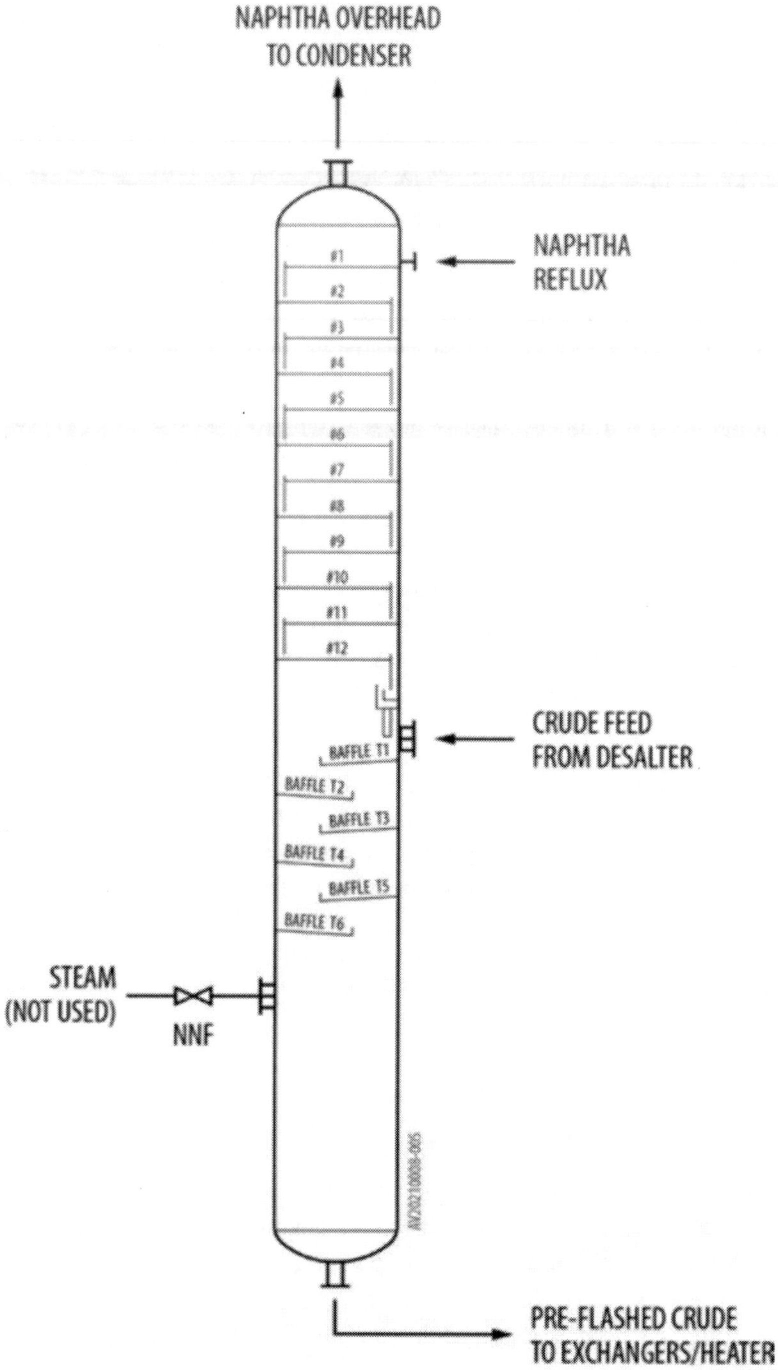

Figure 1. Bazan's Preflash Tower

The tower experienced dark naphtha for many years since the 1990's, about 2 to 3 episodes per year while distilling crudes with high naphtha content. Following a turnaround in 2018, the unit started running on September 9th, 2018. Beginning November 11th, 2018, the preflash tower experienced episodes of dark naphtha product. The dark naphtha episodes happened about 15 times in a year, each dark naphtha event lasting about 12 hours, throughout which the naphtha was continuously dark. The dark naphtha was accompanied with a high pressure drop which remained high throughout the episode. There was no problem of instability of the bottom level during the dark naphtha episodes, but during the episodes the bottom level dropped by approximately 20%.

The rest of the time the tower was working fine because the refinery limited the tower inlet temperature and the total amount of naphtha in the top. To get out of the event, the refinery needed to cut crude charge rates or the feed temperature. These episodes became the prime crude unit capacity bottleneck.

In Search of the Root Cause

Figure 2 shows the bottom seal pans arrangement. The bottom downcomer enters a 242 mm wide primary seal pan, clearing its floor by 67 mm. Liquid from the primary seal pan overflows a 117 mm tall overflow weir into the 386 mm wide secondary seal pan. From the secondary seal pan liquid descends onto the top shed deck via a 4" diameter, 800 mm long pipe. The overflow weir of the secondary seal pan rises 65 mm above the overflow weir of the primary seal pan, so no liquid should be overflowing the secondary seal pan unless the 4" pipe plugs. Figure 2 also shows the 10" crude feed nozzle at an elevation about two thirds of the way down along the 4" pipe.

When the tower was inspected at the 2018 turnaround, the 4" pipe was found broken sitting on the top shed deck. It broke where it exited the secondary seal pan, leaving a crack and 4" hole in the seal pan floor. The seal pan floor was repaired and the pipe reinstated. In the next turnaround, in 2020, the tower was inspected, and again the 4" pipe was found broken at the same spot, leaving a crack and a 4" hole in the secondary seal pan floor (Figure 3).

Figure 4 shows the cause of the pipe breakage. It was located in front of the flashing crude feed inlet, and the high feed velocities caused the pipe to vibrate and break. This breakage was linked to the entrainment experience in the tower. With the pipe broken, the high velocity vapor was theorized to pick up the liquid descending from the secondary seal pan and entrain it up the tower, causing the naphtha to turn dark. A solution was sought to either relocate the 4" pipe or to extend the 10" inlet pipe, with baffling or a pipe distributor, to prevent impingement of the feed on the 4" pipe.

Aware of Fluor's distillation expertise, Bazan invited Fluor to join their task force seeking the root cause for the problem. The joint task force review of the past experience questioned the theory that entrainment of liquid falling from the secondary seal pan is capable of being the root cause for the dark naphtha. The liquid flow rate in the upper trays of preflash towers is small, in many preflash towers too small to keep the downcomers sealed. A 4" pipe does not transport much liquid. As long as the downcomers do not lose their seals, they will drain the liquid entrained from the broken 4" pipe. This small amount will not explain the flooding indicated by the observed pressure drop rise and its persistence for 12 hours. It would take a massive amount of liquid to explain the pressure drop rise and staying high for 12 hours, and this can only come from the incoming crude.

Another important observation is that the upper seal pan was not damaged. So the bottom downcomer will always remain sealed. This would permit the bottom tray to keep operating normally. Since the liquid flow rate in the rectifying trays is small, any entrainment of part of this liquid will be captured by the bottom tray, and will descend via the upper seal pan into the lower seal pan (Figure 2). It will then try to come out through the hole left by the broken pipe. If the rising vapor prevents this liquid from descending through the hole, it will overflow the weir on the secondary seal pan, and again flooding will not occur.

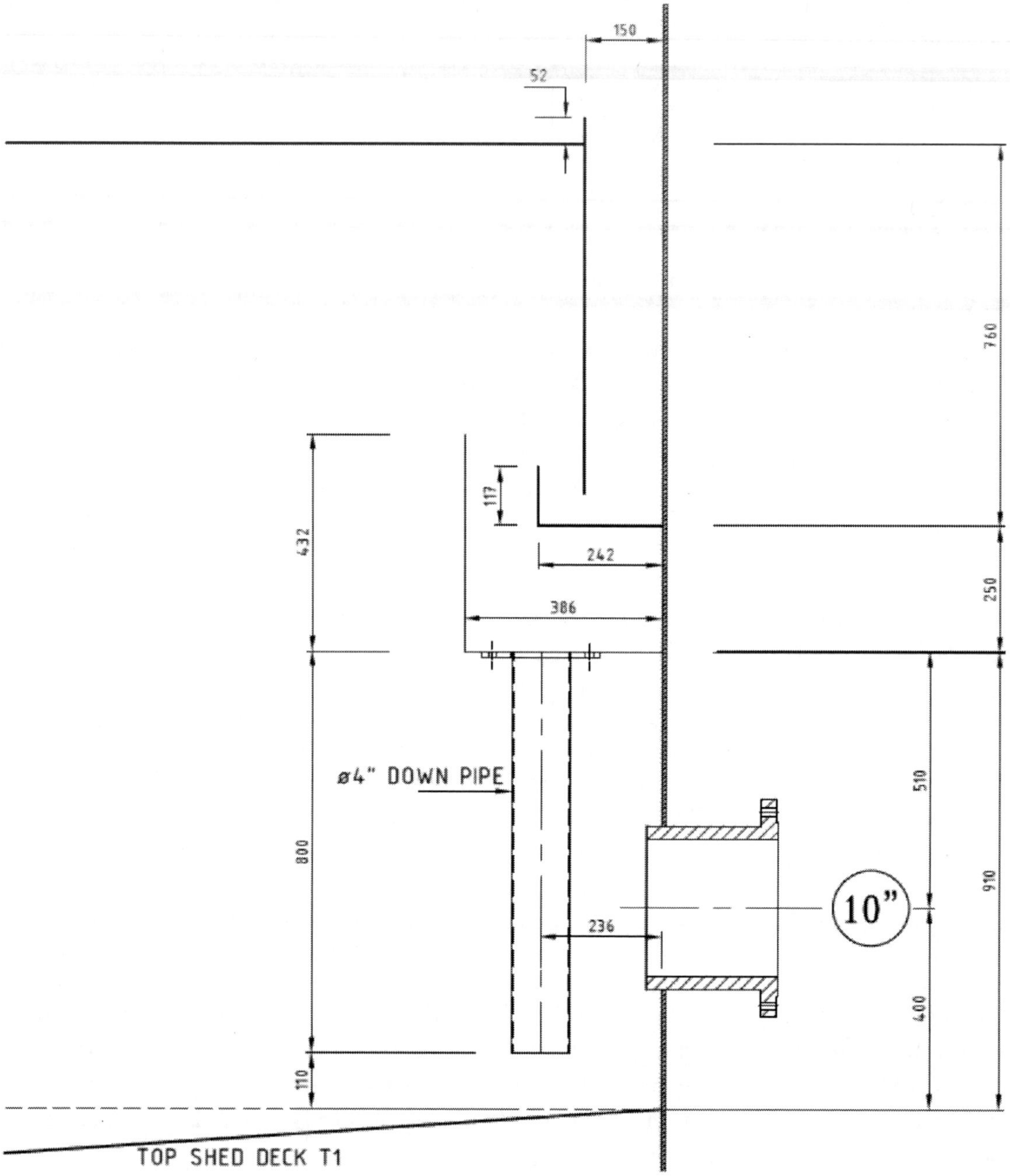

Figure 2. Primary and secondary seal pan arrangement, showing the 4" down pipe from the secondary seal pan to the top shed deck

Figure 3. Broken 4" down pipe and the 4" hole in the secondary seal pan floor

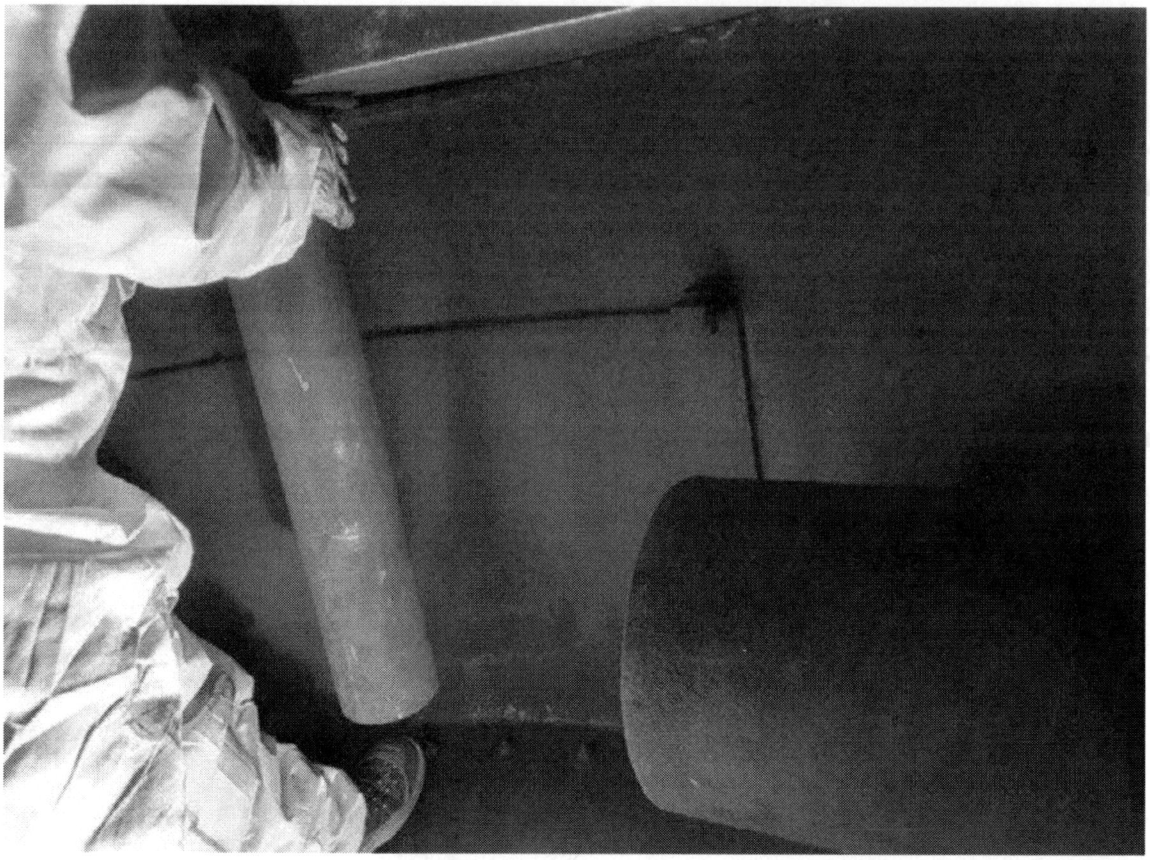

Figure 4. The broken 4" down pipe hand-held in its location, right in front of the tower crude feed inlet pipe

The task force identified two mechanisms by which a massive carryover of crude is likely to occur:

- Preflash towers often experience foaming (1, 2). A foamover usually causes massive carryover of crude to the upper section.
- High $\rho_M V_M^2$ (where ρ_M is the mixed phase density, kg/m^3; and V_M is the mixed phase velocity at the tower entrance, m/s)

Both these mechanisms are consistent with the dark naphtha, high pressure drop and the drop in bottoms level observed. Foaming may be caused by components in the crude (1), desalter chemicals (2), or pockets of water that quickly vaporize upon entering the tower.

A weak argument against the foaming theory is the stability of the tower bottom level. In many cases where preflash foaming occurs, the bottom pump cavitates and the bottom level is unsteady. Neither was observed in the Bazan preflash tower.

The high inlet velocities and foamover mechanisms are interconnected, as has been demonstrated by Barber and Wijn (1), who point out the significant influence of the feed inlet design on preflash foaming. A common practice to cure or prevent foamovers in preflash drums, towers, and similar services is to pass the feed through vortex clusters (3, 4).

With the crude unit coming out of turnaround, adding vortex clusters was not an option. A quick solution was sought.

Hydraulic Analysis and Root Cause

The flashing crude feed entered the tower via a short 10" pipe as seen in Figure 4. The bottom of the 10" crude inlet pipe was only 273 mm above the top shed deck. At the intended operating conditions, the feed mixture was 4.8% vapor by weight. By volume, the mixture was 89% vapor. It was coming in at a huge feed mixture velocity of 16.7 m/s. The $\rho_M V_M^2$ at the 10" pipe and nozzle was 26,600 kg /ms^2. This is about 7 times higher than the common good practice criterion of keeping it below 3,700 kg /ms^2. At this high $\rho_M V_M^2$, which signifies a high kinetic energy, and coming in so close to the shed deck floor, it is highly unlikely that the tower would have worked. The high velocity and kinetic energy would blow the liquid on the top shed deck T1 right up the tower.

Two other factors further aggravated the situation. First, the open area of the top shed deck was low, about 34% of the tower cross section area. By itself this would not have been so bad, as there was only a small amount of vapor coming up through the window from underneath. However, with the feed pipe discharging so close to the floor, the top shed deck was holding a very foamy mix, which would have had difficulty descending via the small window. Second, the liquid velocity down the tower (below the feed entry), based on the superficial area, was 95 m^3/ h m^2. At this rate, the Barber and Wijn correlation (1) predicts a foam height of 1.7 meters. Looking at Figure 2, this foam will almost reach the bottom valve tray.

We concluded that the feed arrangement and discharge velocities were the root cause of the dark naphtha events.

Fix-up Strategy

The corrective action needed was to lower the point of feed discharge into the tower, reduce the kinetic energy of the discharged feed, effectively separate the vapor from the liquid upon entry to prevent foaming, and prevent the feed from impinging on the top shed deck liquid. The conceptual process design of the new feed pipe was performed by Fluor, the mechanical engineering and fabrication by Bazan. Koch-Glitsch, who have vast expertise in tower internals and feed arrangements, joined the task force as consultants on optimizing the feed piping design and on achieving a sound mechanical design. The mechanical design was a special challenge due to the very high kinetic energy of the feed mixture. Neither Fluor nor Koch-Glitsch had any involvement in the original design of this tower and its internals.

The new feed pipe is shown in Figure 5. The 10" feed pipe was extended into the tower, ending in a tee that split the feed into two 10" laterals. The laterals had 93 mm wide slots pointing downwards at 37 degrees to the horizontal (see also Figure 6), discharging the feed onto a new vertical impingement baffle. This impingement baffle directed the liquid downward towards shed deck T2, with the vapor disengaging upwards. To allow free rise of vapor from the impingement baffle, two segments of the top shed deck, T1, were removed. The segment of T1 at the opposite end of the tower from T2 was kept, and was used to collect liquid from the trays that descended from the secondary seal pan via the 4" pipe. To distribute this liquid, a 50 mm weir was installed at the end of the unremoved segment of the top shed deck T1.

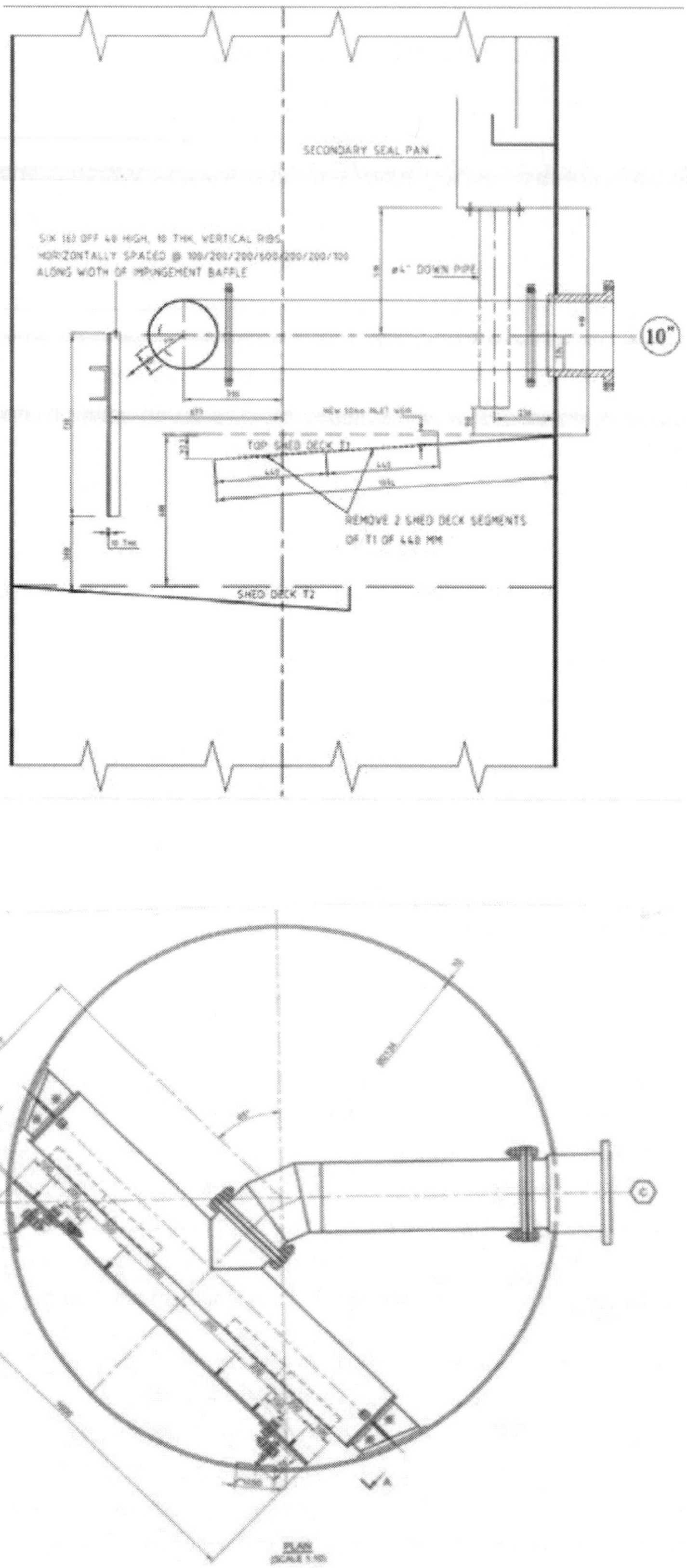

Figure 5. New feed entry design. Top: elevation; bottom: plan

Figure 6. The laterals with the slots that directed the feed towards the impingement baffle at 37 degrees to the horizontal

Taking out the two segments of T1 was central to the success of the design. The vapor rises on the inside of the impingement baffle, so the region at the back of the impingement baffle (between the baffle and the tower shell) does not contribute open area for vapor rise. The vapor velocity through the open area is expressed by the C-factor, given by

$$C = U\sqrt{\frac{\rho_G}{\rho_L - \rho_G}} \quad (1)$$

where C is the capacity factor, m/s, based on the open area for vapor rise; U is the vapor superficial velocity based on the same open area, m/s; ρ_G and ρ_L are the gas and liquid densities, kg/m^3. With no T1 panels removed, the C-factor would have been 0.092 m/s, which is too large for a foaming system. With two panels removed, the C-factor declined several-fold to an acceptable 0.020 m/s.

The new feed entry design faced many challenges. First, before splitting to the tee, the feed mixture velocity in the extended feed pipe was a high 16.7 m/s. The $\rho_M V_M^2$ at the 10" pipe was a huge 26,600 kg /ms^2, about 7 times higher than the common good practice criterion of keeping it below 3,700 kg /ms^2. At this very high $\rho_M V_M^2$, which signifies a high kinetic energy, there would be intense hydraulic pounding of the pipe and a large bending moment. The tee was strongly anchored to the tower shell. The flanges on the feed entry pipe, as well as the impingement plate supports, were double-nutted to have extra protection against vibration and hammering.

In the laterals, the mixture velocity was halved to 8.3 m/s, and the $\rho_M V_M^2$ was 4 times lower, 6660 kg /ms^2. The slot velocity was set roughly the same as the velocity in the laterals, giving the same $\rho_M V_M^2$, 6660 kg /m s^2. Even though this is a lot better than in the 10" main feed pipe, it is still about 80% above the recommended 3700 kg /ms^2. But this is the best that could be done with the tight space constraints in this 2.1 m diameter tower.

Upon turning at the tee, the mixture momentum becomes horizontal, directed towards the blanked end of each lateral. This horizontal momentum will persist after the mixture exits the slots. So the exiting feed mixture will hit the baffle with considerable momentum directed towards the shell ends of the baffle. The large amount of liquid descending in a small region at the end of the baffles near the shell is likely to generate foam upon impact on the T2 shed deck liquid, as well as liquid maldistribution on T2, with high flows near the shell and low flows in the center. The large amount of vapor rising near the shell end of the baffle can generate entrainment. These hydraulic issues are likely to reduce the effectiveness of, possibly even defeat the success of, the new design.

The above scenario can be prevented by installing vertical ribs on the baffle. Figure 7 shows the ribs. The liquid flowing along the baffle towards the shell will be directed downwards by the ribs, and the vapor deflected upwards. This is a patented design (5).

The laterals and the baffle, as installed, are shown on Figure 8. As it was expected that the baffle will receive severe pounding by the feed mixture, it was fabricated from a 10 mm thick plate and anchored strongly to the tower shell.

Epilogue and Takeaways

The tower was restarted on the in 22nd of July 2020. It was in operation until September 6th when the unit was shut down due to falling demand for petroleum products resulting from the Covid-19 pandemic, then restarted on the 20th of November. During the 6 months in operation, no incidents of dark naphtha were experienced. The naphtha production rate was raised by more than 20% without any adverse effects on operation. Bazan did not push yet the unit to the maximum, but intends to do so soon. There was no limit in the preflash tower even at the >20% higher naphtha production. The entrainment problem has been eliminated. The preflash tower is no longer a bottleneck of the crude unit.

A takeaway from this revamp is that the feed inlet design to preflash towers can make or break not only the tower performance, but that of the entire crude unit. Poor design of the inlet can severely bottleneck the tower and the entire crude unit. Judicious, low-cost modifications to the inlet design can largely improve performance, eliminate bottlenecks, and provide trouble-free operation.

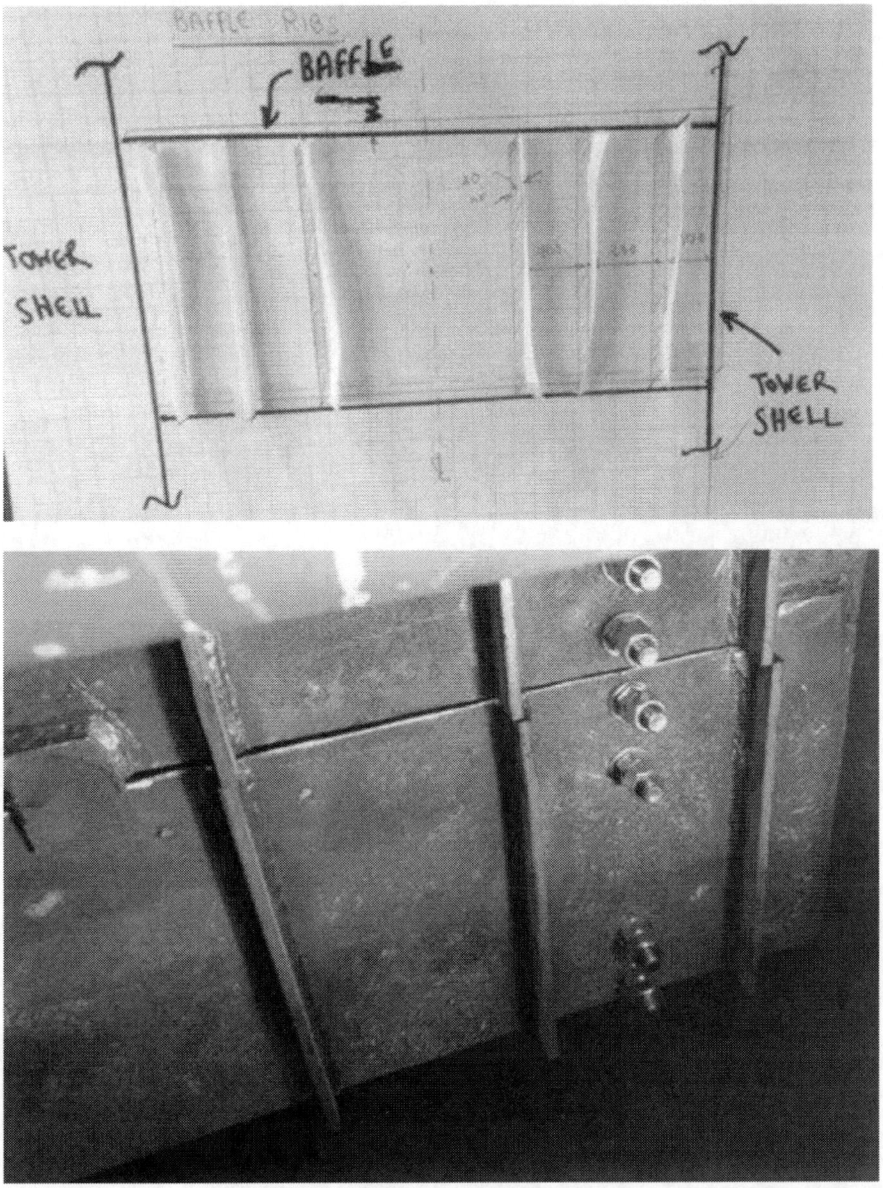

Figure 7. The ribs installed on the impingement baffle to prevent the horizontal momentum of the feed mixture issuing from the slots from pushing the liquid and vapor towards the column shell.

Top: a paper model, used during the design to explain the concept to the fabricators

Bottom: the ribs installed on the baffle

Figure 8. The laterals and the ribbed impingement baffle. The feed pipe entering the tee is obscured by an engineering drawing.

References

1. Barber, A. D., and E. F. Wijn, "Foaming in Crude distillation Units", I.Chem.E. Symp. Ser. 56, p. 3.1/5, 1979.

2. Lieberman, N. P., and S.W. Golden, "Foaming is Leading Cause of Tower Flooding", Oil & Gas Journal, p.45, August 14, 1989.

3. Turner, J., R. J. Asquith, and R. Atkinson, "Stop Foaming on Hydrotreater 'Hot' Separator", hydroc. Proc., June, p. 119, 1999.

4. Barletta, T., E. Hartman, and D. J. Leake, "Foam Control in Crude Units", Petroleum Technology Quarterly, p. 117, Autumn 2004.

5. Kister. H. Z., and W. J. Stupin, "Configurations and Methods for Ribbed Downcomer Wall", US Patent 7,581,719 B2, September 1, 2009.

619640 The MRV-Fixed Valve – A basis for high-capacity solutions

Abstract:

Increasing process intensification in the chemical industry leads to an increasing demand for high capacity mass transfer equipment. Applications are typically driven by minimization of invest for grass-root plants or a required capacity increase in revamp scenarios.

For high capacity mass transfer trays, mini fixed valves are often used as mass transfer units. These elements offer excellent performance concerning maximum gas load and show low pressure drop. Therefore these units are an ideal basis for high capacity solutions.

In the presentation, the performance of RVT's mini fixed valve, the MRV-valve will be shown and discussed. For the systems C6/C7 and iC4/nC4 tests under various pressures were conducted in a 4-foot column. For each system, three different operating scenarios (total reflux, $L/V>1$ and $L/V<1$) were tested. All test runs started at min. load and ended at the flooding point. Efficiency, pressure drop and capacity were recorded; the hydrodynamics on the trays were investigated using gamma scanning. The performance including flood mechanisms is discussed and the results are compared against conventional valves.

The MRV-valve is characterized by excellent performance at high loads and provides a very good and stable efficiency over a wide operating range. These features make it an ideal basis for high capacity trays. It will be shown how the MRV-valve – in combination with e.g. push valves and special downcomer geometries – is used in the design of modern high-capacity trays.

619645 The Uncertainty of Predicting Liquid Entrainment in Trayed Columns

Abstract:

The effects of liquid entrainment create uncertainty during the design and troubleshooting operations of trayed columns. The resulting movement of liquid droplets also reduces countercurrent mass transfer through back-mixing from tray to tray, as well as along the flow path on a tray deck. The carryover of liquid to trays above also leads to jet flood with increasing vapor rates. The prediction of entrainment still leaves an engineer on tenuous ground even though the mechanisms related to tray design and physical properties have been studied in detail. This presentation will discuss insights into the impacts and prediction of entrainment phenomena though research completed over the past 70 years.

Feed Entry into Parting Boxes: What Works and What Does Not

By

Henry Z. Kister, Garry Jacobs, and Abraham A. Kister

Fluor, Aliso Viejo, CA, USA

Presented at the Distillation Symposium, AIChE Spring Meeting, Virtual, April, 2021

Feed Entry into Parting Boxes: What Works and What Does Not

By Henry Kister, Garry Jacobs, and Abraham A. Kister, Fluor, Aliso Viejo, CA

Achieving good liquid distribution is crucial for the performance of packed towers and multi-pass tray towers. With the focus on liquid distribution, one device that has received surprisingly little attention in the published literature is the parting box. While good-practice guidelines are available, the published literature is short of work on the hydraulics of parting boxes and tested guidelines for its design.

To gain insight into the parting box performance, we built a scaled-down model, geometrically similar to the parting box in a hydrocarbon tower, and tested it with water. This model and the tests were intended not only to shed light on the tower issue, but also to provide generic insight into parting box hydraulics as tested in a realistic parting box design.

Our earlier tests[1] showed that when feeding from a sparger pipe, the liquid in the test parting box was highly frothed at both high and low liquid rates. The frothing was due to water jets from the feed pipe entraining air into the water upon impact. This generated excessive froth heights in the parting box, and premature overflow at high rates. The liquid split to the five clusters of holes at the floor of the parting box was good at the maximum rates but quickly deteriorated as the liquid rate was turned down. Later tests[2] showed that lower hole velocities out of sparger pipe lowered froth heights and somewhat reduced aeration in the parting box, but the liquid split was again good only at the maximum rate, deteriorating upon turndown.

Adding a Perforated Impingement Plate (PIP) with either 20% or 10% open area, mounted either 2" or 4" above the parting box floor, greatly improved the split of liquid to the clusters of holes at the floor of the parting box[1]. With a PIP, the split was very good to excellent both at the maximum rates and at turndown down to about 50% of the maximum rates. Lower rates were not tested.

In the tests with PIP's, an air gap formed beneath the PIP, with tall froth above, producing premature overflows at high rates. The air gap resulted from the water jets issuing from the feed pipe entraining air with them. This air descended through the PIP, became trapped by the prevailing water head on top of the PIP, and pressured up the space between the parting box floor and the PIP. In the current tests, we tested narrow perforated plate designs which provided much better venting.

Our previous work[2] also tested a design in which dip tubes were used to discharge the incoming liquid 1.75" above the floor of the parting box. This fully eliminated the frothing at all but the lowest rate, but again the liquid heights in the parting box were considerably higher than those calculated from the orifice equation. We hypothesized that the horizontal momentum generated upon sideways deflection of the liquid by the floor of the parting box made it more difficult for the liquid to descend through the parting box holes, causing larger liquid heads. At the lower rates, the dip tubes also appeared to cause air to be sucked into the sparger pipe via its vent holes. In the current tests, we repeated these tests with larger-diameter dip tubes (reduced liquid velocities).

Our previous work[2] identified one satisfactory method of feeding the parting box. This method used dip tubes, terminating either 1.75 inches or 5.5 inches above the parting box floor, which were fitted with tees that directed the liquid towards the parting box walls. This method was shown to have eliminated the excessive liquid height problem, gave good split of liquid to the clusters of holes in the floor of the parting box, and a head-flow relationship closely matching the orifice equation.

Our current work seeks additional satisfactory methods for feeding parting boxes. The success of the tees at the end of the tubes suggested to us that dispersing the incoming momentum of the entering liquid jets is key to good feeding of parting boxes, and we tested a number of designs that promised to achieve this.

Another question addressed in the current tests is the impact of the orientation of the main liquid supply pipe into the feed sparger tee. In all previous tests, the liquid supply pipe entered perpendicular to the tee in the middle of the sparger. In practice, the supply pipe often enters at an angle, so this time we performed a number of tests with the feed pipe oriented 45° to the sparger pipe.

Test Rig and Procedure

The current tests were performed using the test rig reported in our previous work[1, 2]. The only change was the feed piping. Our test rig is a scaled-down model, geometrically similar to the parting box in a hydrocarbon tower.

Figure 1 shows a top view of the test rig. The parting box was rectangular acrylic 21.5" long x 2.75" wide x 9.75" tall, and contained a perforated floor plate 20.5" x 1.75" cemented to the bottom flange. The floor plate had 75 holes at 3/16" diameter drilled into it using a press drill, in five clusters as shown in Figure 1, mimicking the hole geometry and arrangement in the floor of the hydrocarbon tower parting box. The total perforated area was 5.8% of the floor plate. The top of the rig was saddle-shaped to support the 2.07" ID transparent PVC perforated pipe sparger.

The sparger, shown in Figure 1 and used in our previous tests[1] and many current tests, had 12 downward-pointing 3/8" holes, six equally-spaced on each side of the tee, with the bottom of the pipe 13" above the parting box floor. Round hose clamps were used to fix the pipe along the parting box length so the center lines of the tee lined up directly above the center lines of the floor plate. Two 5/32" vent holes, each at the opposite end of the tee, were drilled at the top of the pipe. Loose round hose clamps were slip-fitted around the pipe and moved over the vent holes to cover them after the air vented out at the beginning of each run.

Tap water to the rig was supplied by a 40 gpm submerged centrifugal pump from the bottom of a 96 gallon drum. A throttling valve in the pump discharge line to the sparger enabled manually adjusting the flow rate. The valve was set at a fixed throttle right at the start of the run to give the desired flow rate. The water level on the drum was kept constant within plus or minus a foot by pumping water into it from two other 64 gallon drums, using two additional submerged pumps. The tests were conducted in the open air in dry weather with ambient temperatures ranging from about 65°F to 85°F.

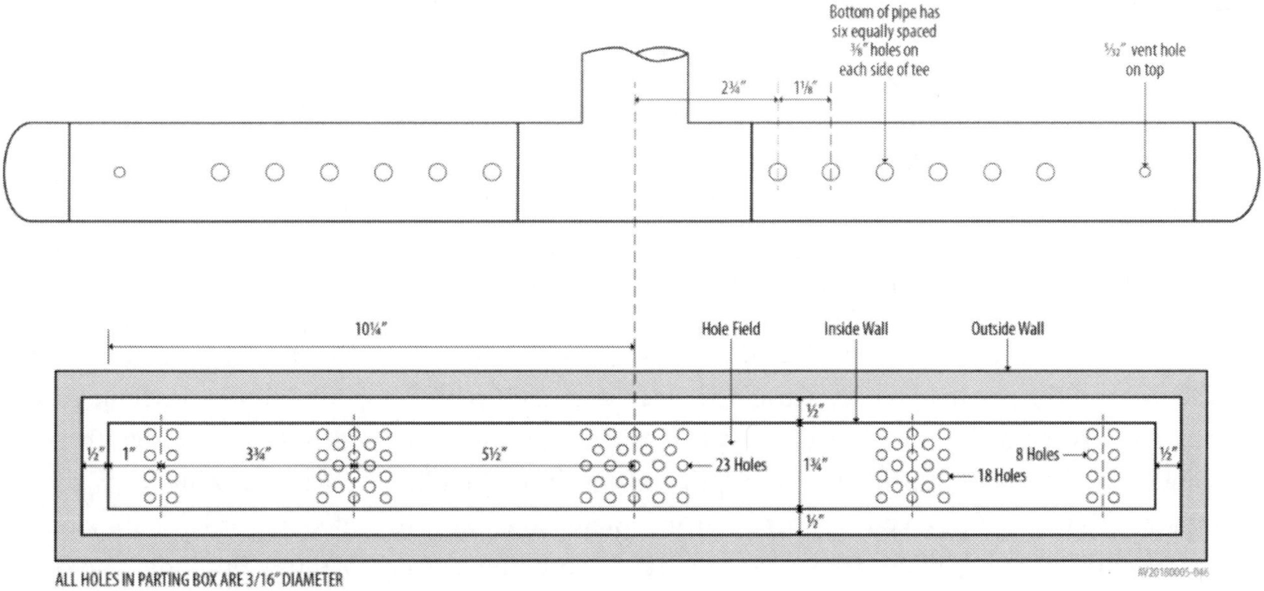

Figure 1. Test Rig Sketch: perforated pipe and parting box floor plans

Water leaving each cluster of holes in the floor plate descended into a separate chute, each diverting the collected water into a separate collector bin by gravity via a flexible pipe containing bellows. Each bin contained a ruler vertically fixed to the bin wall. At the end of each run a photograph was taken of the liquid level at the ruler in each bin. Prior to the tests, and again at the end of the tests, each bin was calibrated by stepwise introducing measured quantities of water and recording the water height in the bin. At the end of the tests the calibration was repeated, this time by stepwise removing the water from each bin by fixed quantities and again recording the heights in each bin after each step. The differences between the two calibrations were small in all the bins. Typically, in the larger bins collecting the liquid from the central three clusters, 1 inch of height corresponded to about 3 gallons, while in the smaller bins collecting liquid from the side clusters 1 inch of height coincided with about 2 gallons.

Each test was concluded after about 110-140 gallons were collected and the central bin approached its maximum desired height. Upon a signal, the pump was stopped and all the bellowed pipes were simultaneously turned upwards so the water remaining in the rig was trapped by the pipes and no longer entered the bins. After all the bin water height readings were completed, the water was pumped back to the feed drums for the next run. At the end of the tests, the water was pumped into garden irrigation hoses so it was all reused and none wasted.

Devices Tested

A number of devices were investigated in the current tests:

Vented Impingement Plate (VIP)

The peripheral ½" wide flanges surrounding the hole field of the floor plate in Figure 1 were used for mounting two peripheral hollow rectangular inserts, one 4" tall, the other 2" tall. In previous tests they were mounted one above the other, sandwiching a flat momentum-breaking plate, referred to as a Perforated Impingement Plate (PIP), between them. Such a plate is used in many commercial parting boxes to arrest the impact of the liquid jets issuing from the sparger on the liquid surface in the parting box. Two PIP's were used in the past tests, one at a time, each containing 3/16" perforations, one with 10% open area, the other with 20% open area. The inserts permitted each PIP to be mounted either 2" or 4" above the floor.

Our previous tests showed[1] that while the PIP's improved the split of liquid to the clusters of holes at the floor of the parting box, they trapped air underneath, which impeded liquid descent, causing excessive froth heights in the parting box. In the current tests, we reduced the width of the impingement plates to 1.5 inches, and reduced the insert width to 7/32", forming gaps of total width of 13/16" for gas venting between the plate and the box walls. Figure 2 compares the previous Perforated Impingement Plates (PIP's) with the current Vented Impingement Plates (VIP's). As in previous tests, two plates were used in the tests, one at a time, each containing 3/16" perforations, one with 10% open area, the other with 20% open area. The VIP's were mounted 4" above the parting box floor. They were supported by short strips resting on the lower insert, positioned such that they did not obstruct any of the PIP open area (Figure 3). No upper insert was used in the current tests.

With the 10% open area plate, we also tested a third VIP plate design, which was a VIP with a piece cut out in the center, right under the unperforated tee of the sparger pipe, that allowed additional venting (Extended Vent Impingement Plate, EVIP, Figure 3). The total unblocked areas at the VIP's, including the plate hole area and the area of gaps between the VIP and the walls, was 48% and 42% for the 20% and 10% open hole area VIP's in Figure 2, respectively, increasing to 46% for the EVIP in Figure 3.

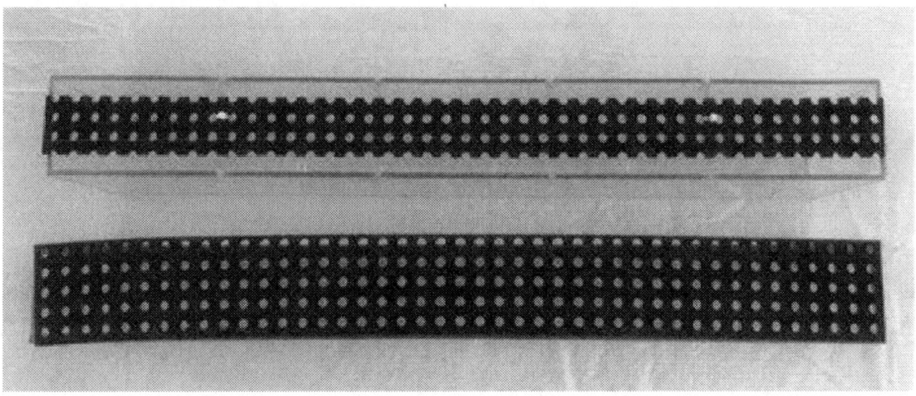

Figure 2. Impingement plates used in the parting box tests, with 20% of plate area perforated.

Upper plate: 1.5" wide Vented Impingement Plate (VIP). The venting areas are the gaps between the walls of the parting box insert and the plate. This type of plate was used in the current tests

Lower: Perforated Impingement Plate (PIP) extending in width from wall to wall in the parting box. This type of plate was used in previous tests[1,2].

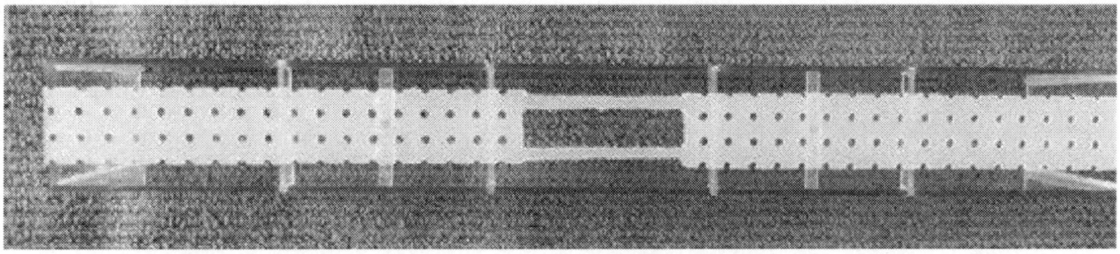

Figure 3. 1.5" wide Extended Venting Impingement Plate (EVIP), in which a central cutout provides additional venting, used in some parting box tests with 10% of plate area perforated. Also shown are the support strips used for supporting the VIP's and EVIP on the parting box insert.

Large-Diameter Dip Tubes

To prevent frothing, many parting box designs use dip tubes that bring the liquid below the parting box liquid surface, often terminating a short distance above the floor of the parting box. This is a more expensive design, often considered ideal.

For the dip tube tests, the perforated sparger pipe in Figure 1 was replaced by a sparger pipe with four dip tubes inserted into the sparger pipe without restriction (i.e., full ID). Each dip tube discharged the liquid midway between every two adjacent clusters of parting box holes. In our previous tests[2] we used four 0.622" ID tubes (½" NPS Sch. 40, NPS abbreviation for Nominal Pipe Size), making the tubes' total cross section area roughly equivalent to the total hole area of the twelve 3/8" sparger holes (Figure 1), which was our base case. The tubes terminated about 1.75 inches above the parting box floor.

Our previous tests with these dip tubes[2] gave excessive liquid heights at the maximum liquid rates, 30% above those calculated from the orifice equation. We hypothesized that the horizontal momentum caused by the liquid jets hitting the floor and being deflected sideways made it more difficult for the liquid to descend through the parting box holes. This raised the question of whether larger dip tubes terminating a little further from the parting box floor will lower liquid height by reducing the horizontal momentum effect. Therefore, the current tests used larger 0.824" ID tubes (¾" NPS Sch. 40) terminating 2.125" above the parting box floor, thus reducing the velocity of liquid exiting the tubes. Table 1 compares the tube velocities to those in the smaller tubes, and in the perforated pipe tests with 3/8" inch holes. The tube velocities in the ¾" NPS tubes were well within the recommended practice guidelines.

Table 1. Calculated Pipe, Hole, and Dip Tube Velocities

Liquid Rate, GPM	27.00	25.00	20.00	12.00	Recommended guidelines (see earlier)
Pipe velocity, ft/s	1.29	1.20	0.96	0.57	Hole velocity>>pipe velocity
Pipe rho vee squared, lb/ft.s^2	104	89	57	20	
Total number of pipe perforations	12	12	12	12	
Pipe perforation diameter, in	0.375	0.375	0.375	0.375	
Pipe perforation velocity, ft/s	6.54	6.05	4.84	2.90	<10 ft/s, desired <4-5 ft/s
Pipe perforation rho vee squared, lb/ft.s^2	2667	2286	1463	527	
½" NPS Dip Tubes					
Total number of dip tubes	4	4	4	4	
Dip tube diameter, in	0.622	0.622	0.622	0.622	
Dip tube velocity, ft/s	7.13	6.60	5.28	3.17	<10 ft/s, desired <4-5 ft/s
Dip tube rho vee squared, lb/ft.s^2	3171	2719	1740	626	
¾" NPS Dip Tubes					
Total number of dip tubes	4	4	4	4	
Dip tube diameter, in	0.824	0.824	0.824	0.824	
Dip tube velocity, ft/s	4.06	3.76	3.01	1.81	<10 ft/s, desired <4-5 ft/s
Dip tube rho vee squared, lb/ft.s^2	3171	2719	1740	626	

Dip (Guide) Tubes with Perforated Tips

When the dip tubes are shortened, discharging a greater distance above the parting box floor, they are submerged in the liquid only at maximum rates, so they are often referred to as "guide tubes". However, to be consistent, we will refer to all as "dip tubes". We previously tested[2] dip tubes terminating 5.5" above the floor of the parting box. These tests gave mediocre split of liquid to the clusters of holes at the floor of the parting box and excessive liquid/froth heights in the parting box. We theorized that the downward momentum of the liquid jets which created interference at the base of the parting box could be the cause. To improve, in the current tests we fitted perforated tips (Figure 4) at the ends of the dip tubes. Our tests used both 20% and 40% open area tips on the smaller dip tubes (½" NPS), terminating 5" above the parting box floor.

Figure 4 Tips mounted below the dip tubes, 20% open area (left), 40% open area (right).

Dip Tubes with Spray Nozzles

In our previous tests[2], we saw excellent performance when we mounted tees at the end of the dip tubes with openings directed towards the parting box walls. This raised the question of whether using sideways-discharging spray nozzles (Figure 5) will give the same result. We tested Spraying Systems Co. ¾" K-120 nozzles that, at the maximum flow rate of 27 gpm, would have a pressure drop of 3 psi, which our pumping system can handle. The four spray nozzles were mounted at the bottom of the dip tubes, with the bottom of the nozzles about 4" above the bottom of the parting box.

Figure 5. Sideways directed spray nozzle used in our tests

Orientation of Main Feed Pipe to Feed Sparger

In all our previous tests, the water supply pipe was perpendicular to the tee in the middle of the feed sparger pipe. In plants, the feed pipe often enters the sparger pipe at an angle. As a result, a question that was brought up is "What impact does the orientation of the water supply pipe to the feed sparger pipe have?" In order to test this, we performed a number of tests with the water supply pipe oriented 45° to the sparger pipe (Figure 6).

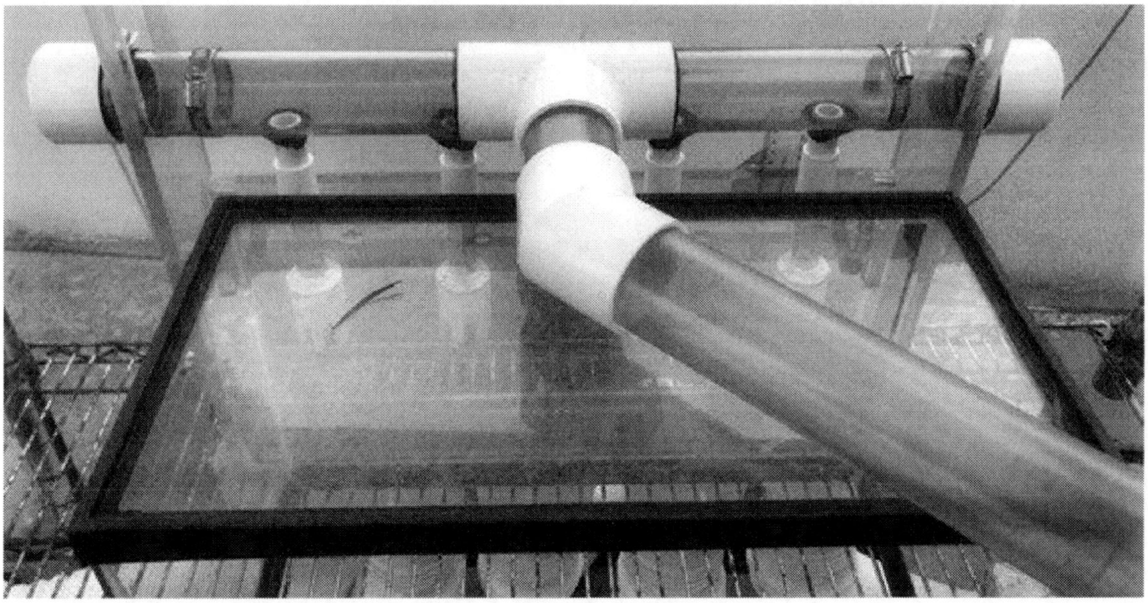

Figure 6. Water supply pipe oriented 45° to the feed sparger

Test Results with Vented Impingement Plates (VIP's)

Figure 7 compares photographs of the current tests with the VIP (vented) to previous tests[1] with a PIP (unvented) using 10% open area plates mounted 4 inches above the parting box floor. The photographs show that the air gaps previously observed under the perforated plates completely disappeared with the vented VIP, leading to lower froth heights.

Figure 8 shows a very good to excellent split of the liquid to the bins with the 10% open area VIP. There are some variations at the lower rates, but they appear small and random. The same seems to apply regardless of the additional venting cutout (shown in Figure 3) in the center of the plate

Figure 9 shows an excellent split of the liquid to the bins with the 20% open area VIP. The same seems to apply regardless of whether the main water supply line entered the sparger pipe perpendicularly or at 45° angle.

Figure 10 compares the froth heights as determined visually from the photographs to the liquid heads calculated from the orifice equation (with coefficients per Perry's Handbook[3]). At liquid flow rates up to 21 gpm, the measured froth heights were close to the liquid heads calculated from the orifice equation. At liquid rates of 29-31 gpm, the froth heights were about 10-20% higher than the calculated liquid heads. In the comparable tests with the unvented PIP[1], the froth heights at 25-27 gpm exceeded the calculated liquid heads by 25-30%. It may be that at the very high liquid rates there was so much air dragged down by the water jets that even the 40-50% VIP vent area could not cope with it, causing an increase in froth height.

In summary, the tests showed that using vented impingement plates (VIP's) with 10 to 20 percent open area, with a vapor rise area of 40-50% of the parting box cross section, gave very good to excellent liquid split out of the parting box. At up to 70% of the maximum rates the measured froth heights were close to the liquid heads calculated from the orifice equation, but near the maximum rates they exceeded those calculated from the orifice equation by 10-20%.

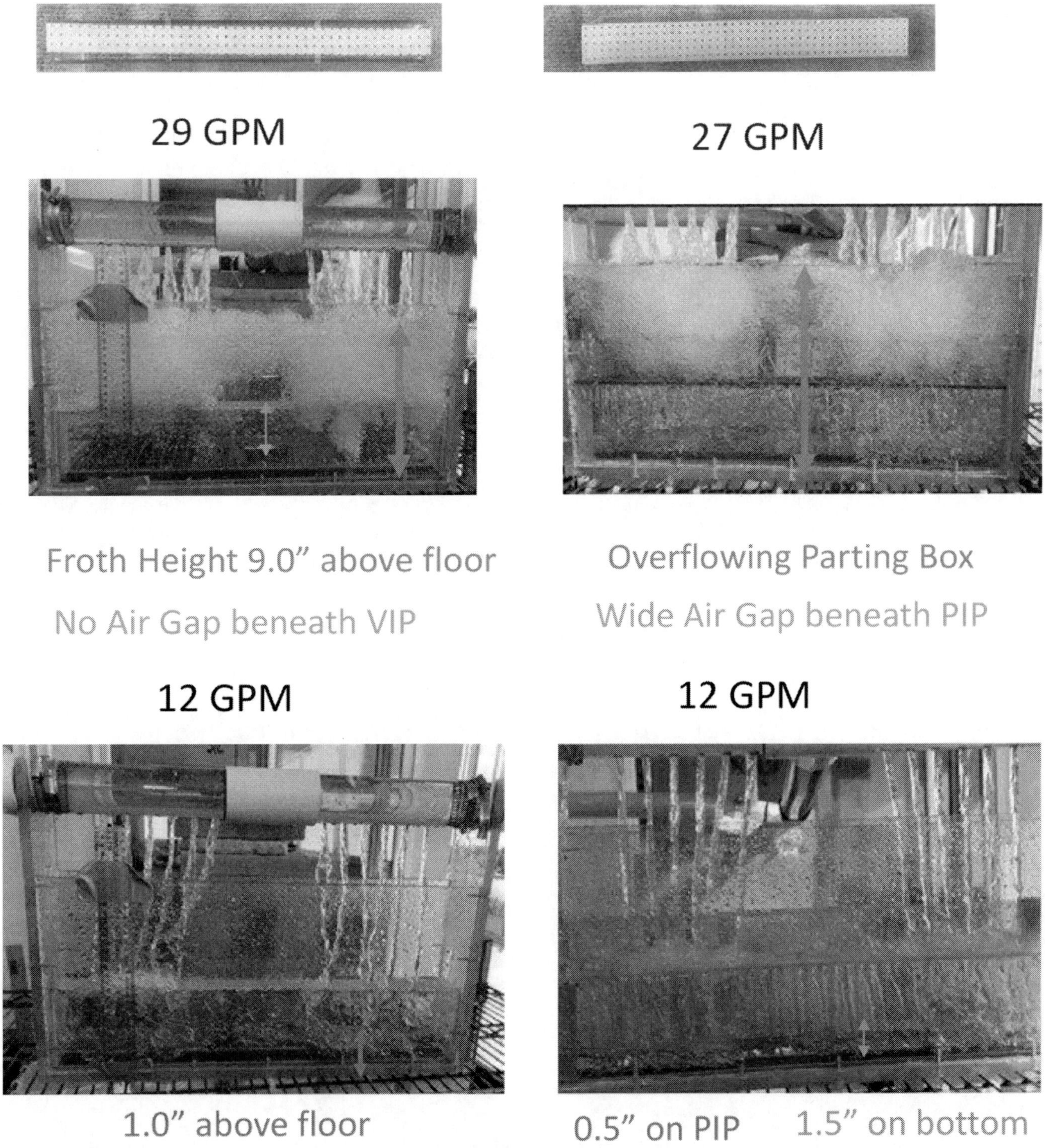

Figure 7. Left: current tests with the 1.5" wide Vented Impingement Plate (VIP), leaving a gap near the walls. Right: previous tests with the unvented wall-to-wall Perforated Impingement Plate (PIP), occupying the total parting box width. In each test, the plate had 10% open area.

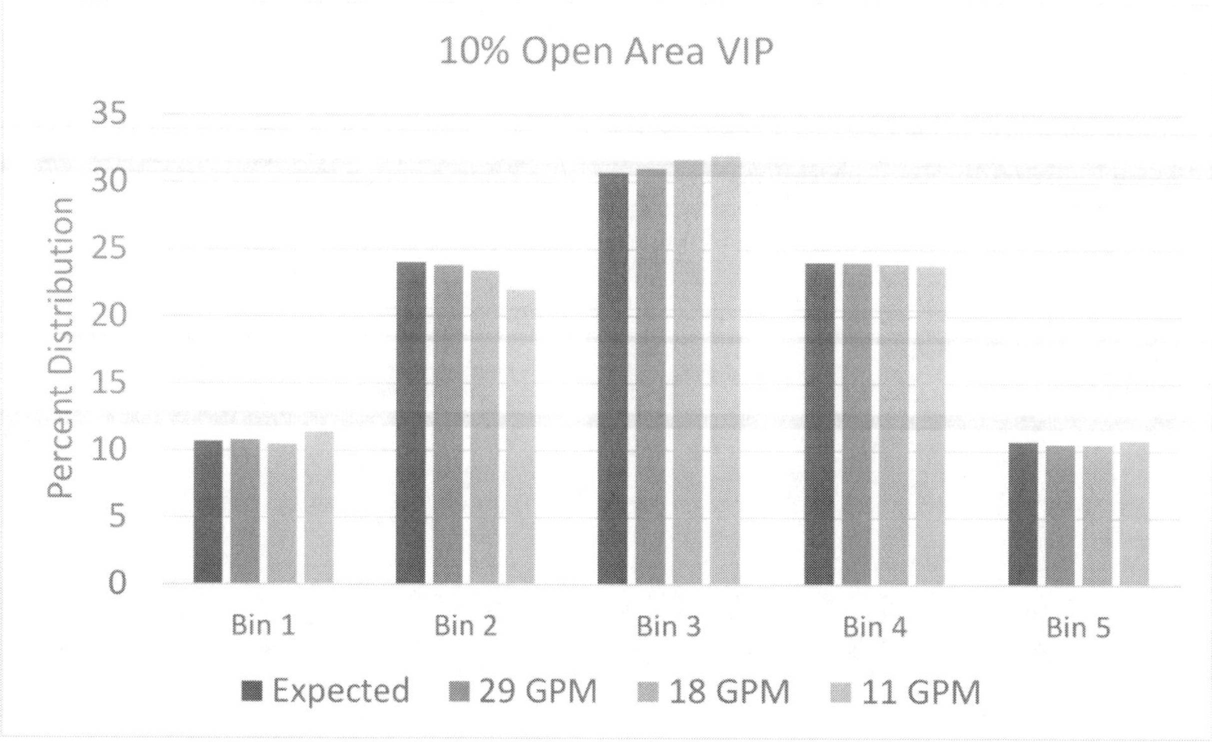

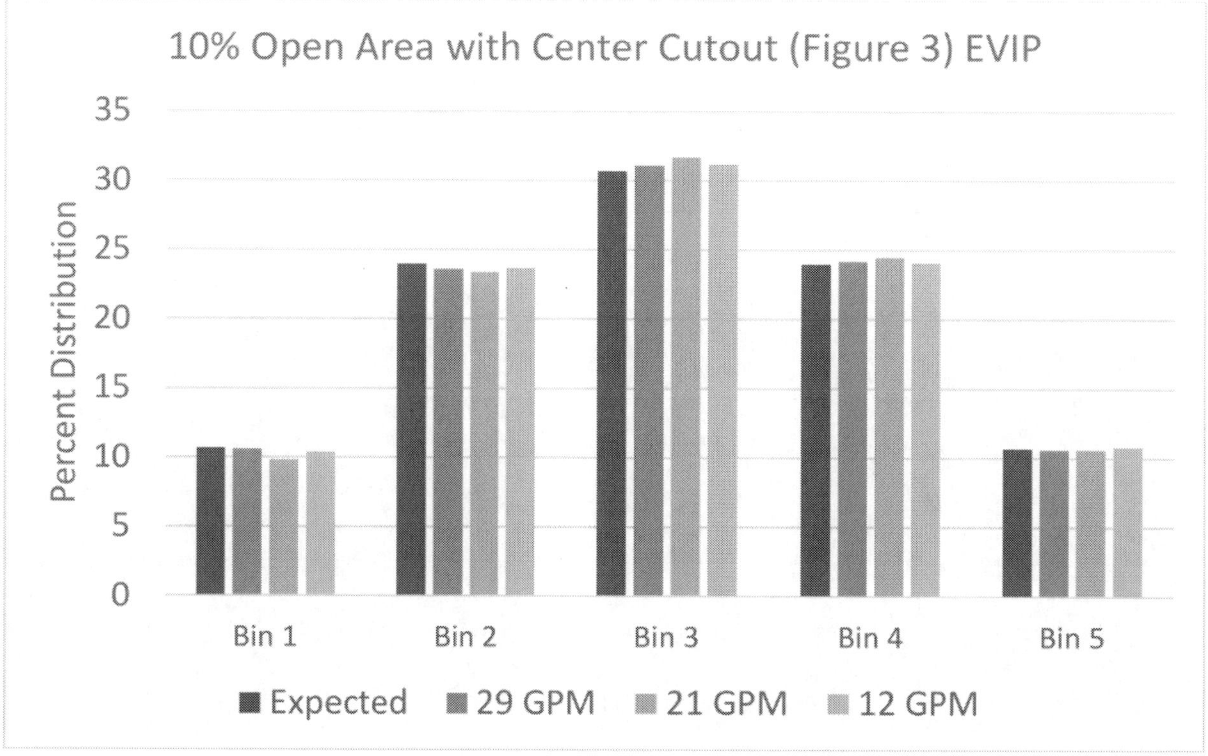

Figure 8. Liquid split among the bins compared to the desired split based of equal liquid flow rate through each perforation. VIP's with 10% open area. Upper diagram, without an additional cutout in the center; lower diagram with additional cutout.

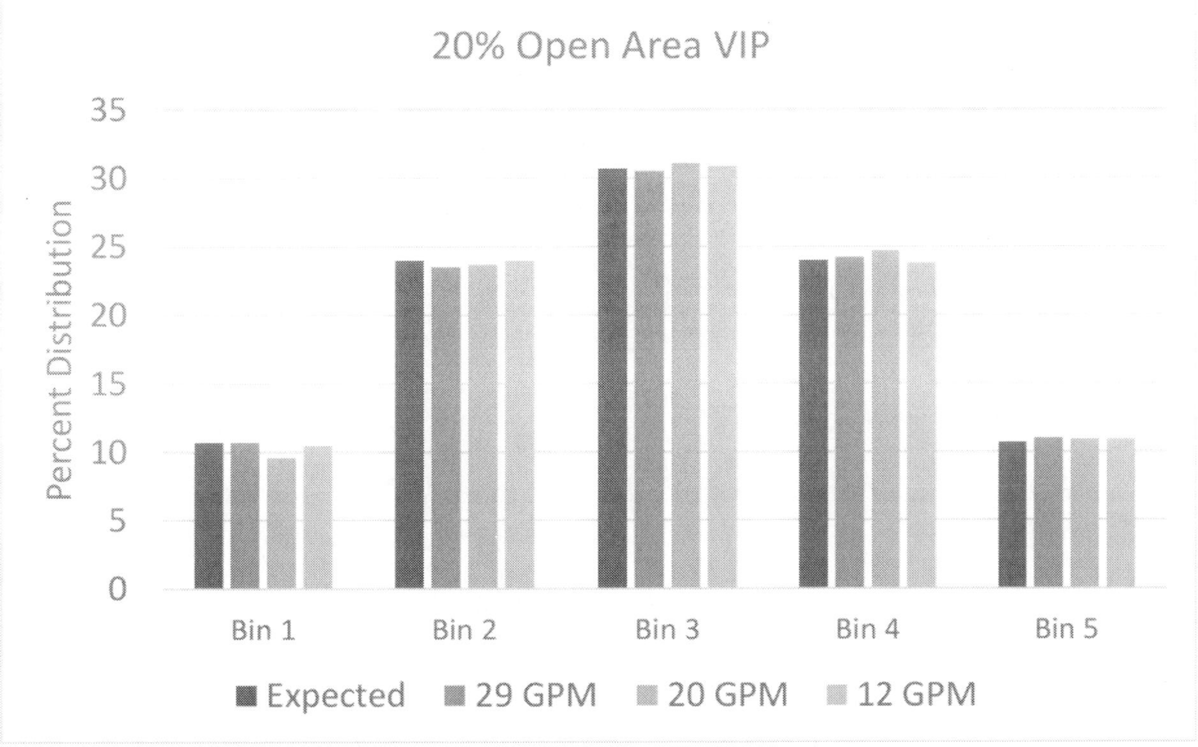

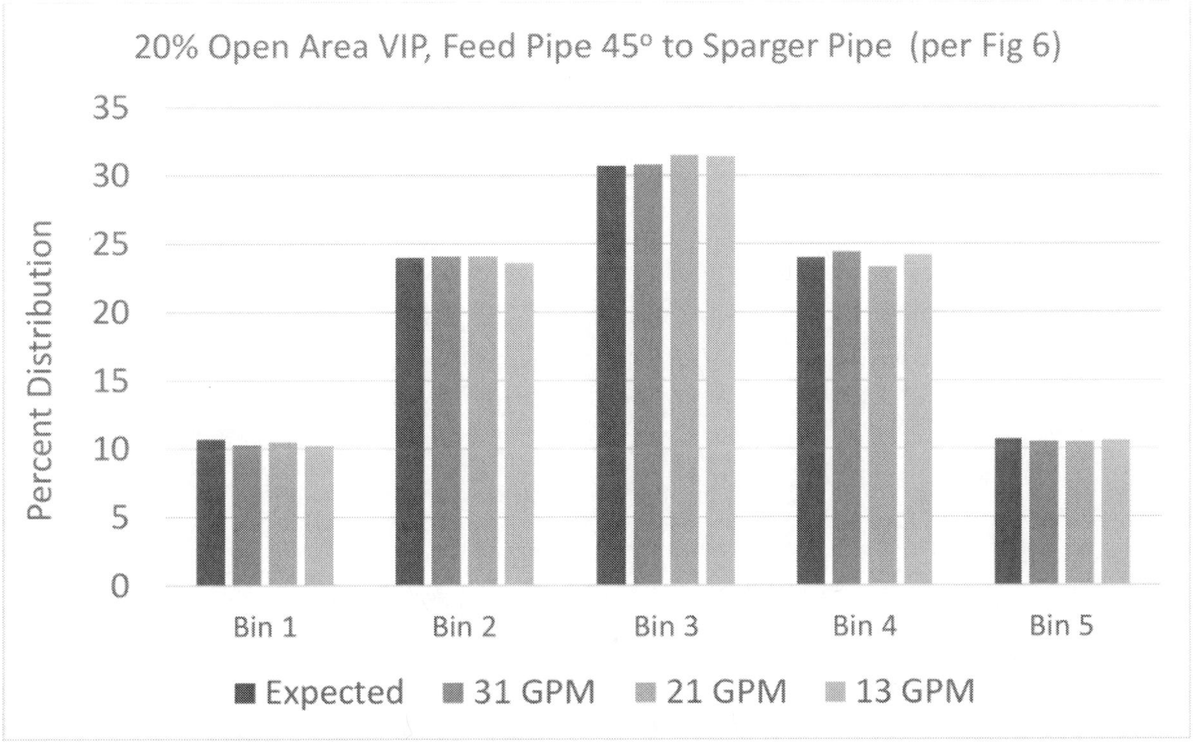

Figure 9. Liquid split among the bins compared to the desired split based of equal liquid flow rate through each perforation. VIP's with 20% open areas. Upper diagram has the water supply line enter perpendicular to the sparger pipe, lower diagram at 45° to the sparger pipe (as in Figure 6).

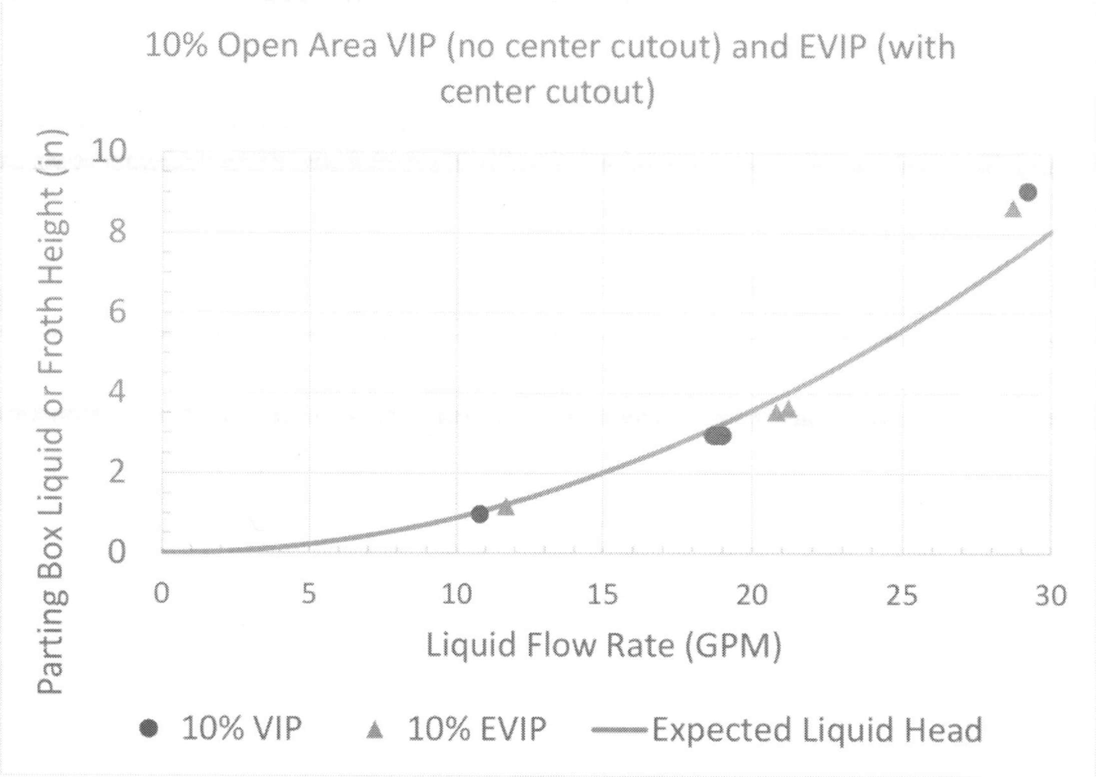

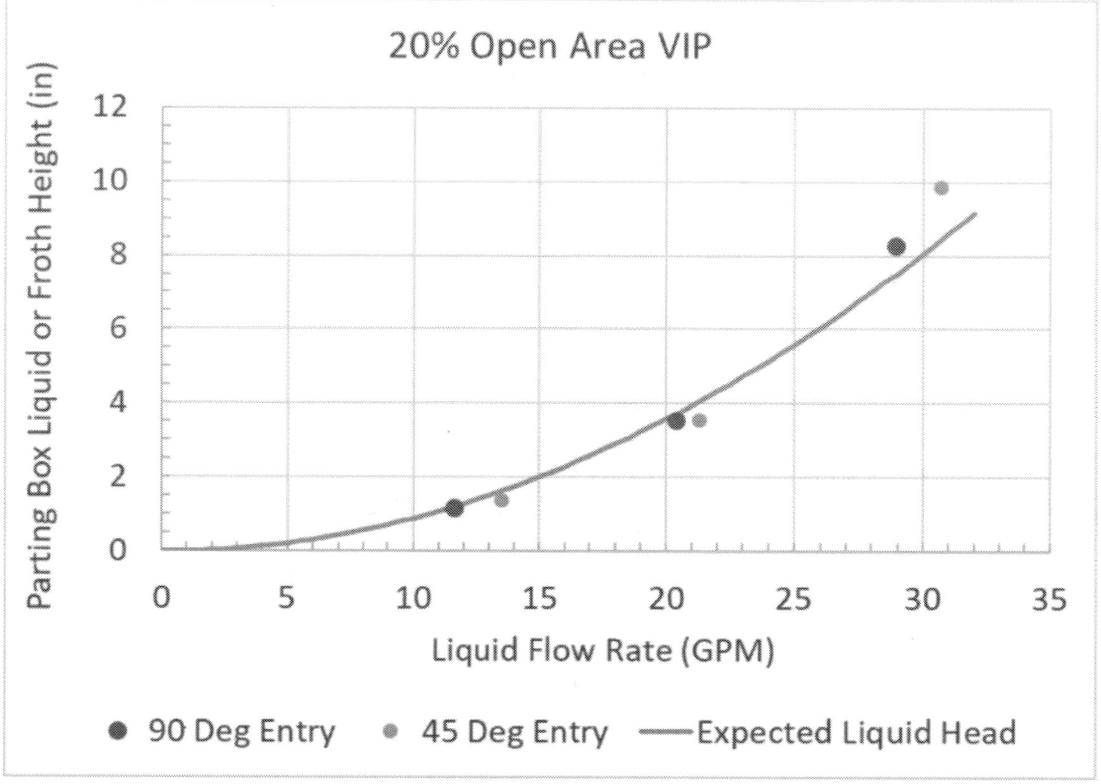

Figure 10. Head- flow relationship, with 10% VIP's and EVIP and 20% VIP. For the 20% VIP, data for water supply line entering at 90° and 45° to the sparger pipe

Test Results with Large-Diameter Dip Tubes Terminating 2⅛" above Floor

Figure 11 shows photographs of the tests with the larger ¾" NPS dip tubes terminating 2⅛" above the parting box floor. For the 30 gpm run, the sparger feed pipe was liquid-full and there was very little frothing in the parting box. At the 20 gpm run and to a greater extent at the 12 gpm runs, the sparger feed pipe was not liquid-full, and air from the sparger feed pipe was seen being sucked into the dip tubes. It appears that the descending water column in the tubes sucked some air into the feed sparger. The air exited at the bottom of the dip tubes into the parting box liquid and aerated it. In the 12 gpm run, the amount of air sucked into each individual dip tube cycled, with some tubes going almost dry in parts of the cycle. Sucking air from the sparger feed pipe was observed to a lesser extent in the smaller (½" NPS) dip tubes[2] at the lower test rates, and is likely to also exist in commercial spargers with dip tubes.

Figure 12 shows an excellent split of the liquid to the bins at the 30 gpm run, a good liquid split at the 12 gpm run, and not so good at the 20 gpm run. In all runs, more liquid issued from the middle cluster and less through the sides and/or off-center clusters.

Figure 13 compares the liquid heights as determined visually from the photographs to those calculated from the orifice equation (with coefficients per Perry's Handbook[3]). Also shown in this figure are previous measurements with ½" NPS dip tubes. At liquid flow rates 20 gpm and below, the liquid height in the box for the ¾" NPS tubes was well-predicted by the orifice equation. At about 30 gpm, the measured liquid heights with the ¾" NPS tubes were about 15-20% higher than the calculated liquid heads predicted by the orifice equation. This rise, to an even larger extent (Figure 13), was also observed in the previous tests with the ½" NPS dip tubes. For liquid rates exceeding 20 gpm, frothing at the parting box was non-existent, so aeration could not explain the difference between the measurement and calculation. The most likely explanation[2] is that the horizontal momentum of liquid issuing near the parting box floor made it more difficult for liquid to descend through the parting box holes, thereby incurring the additional liquid head. With the larger dip tubes the liquid momentum is reduced and the rise in parting box liquid became less pronounced.

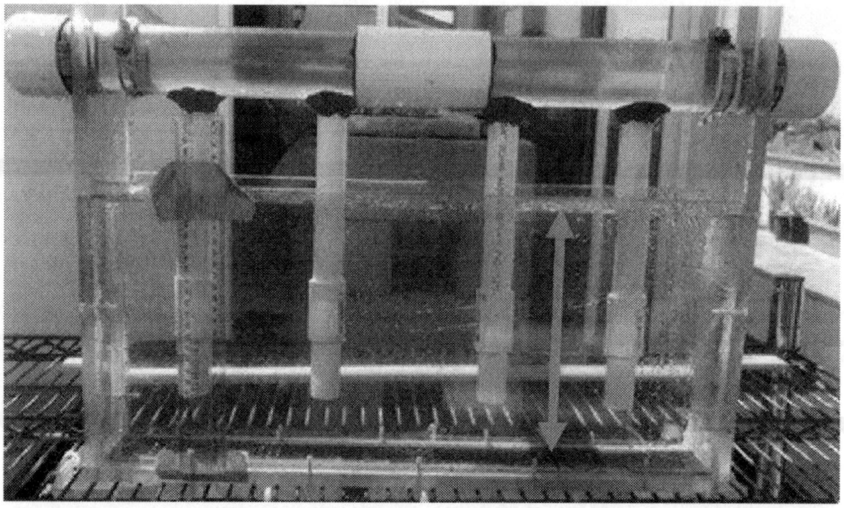

30 GPM

9.4" above floor

20 GPM

3.5" above floor

12 GPM

1.4" above floor

Figure 11. Tests with ¾" NPS dip tubes terminating 2.125" above the parting box floor,

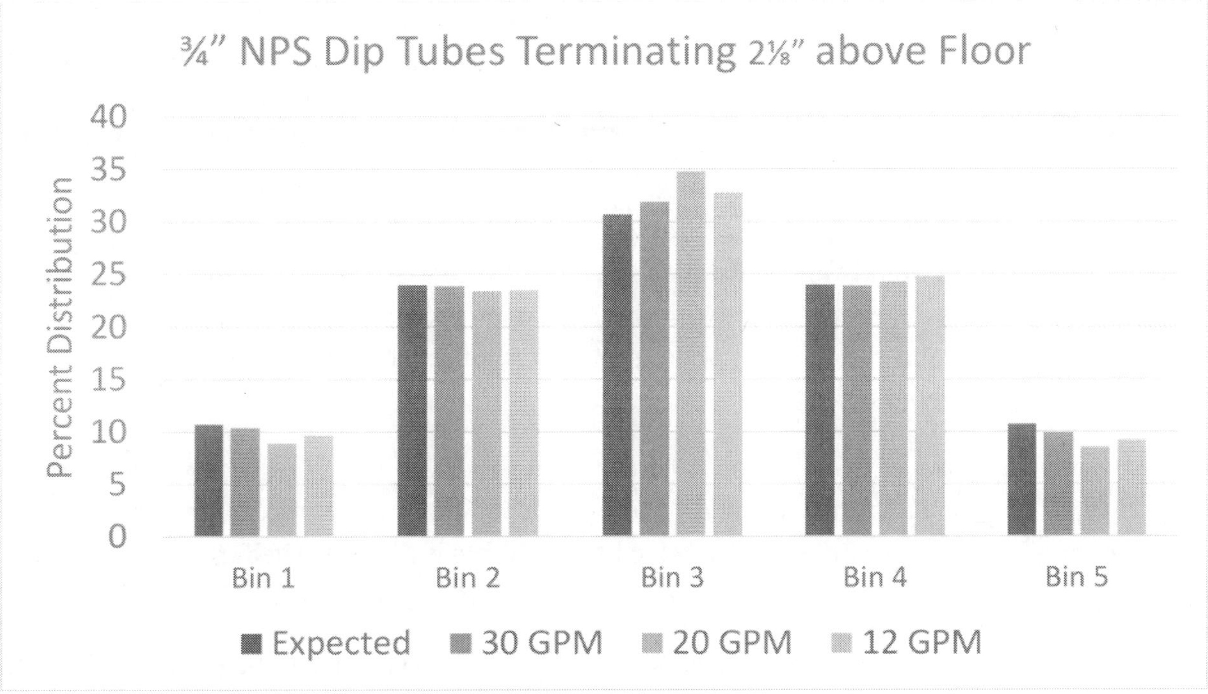

Figure 12. Liquid split among the bins compared to the desired split based of equal liquid flow rate through each perforation. ¾" NPS dip tubes terminating 2⅛" above the parting box floor.

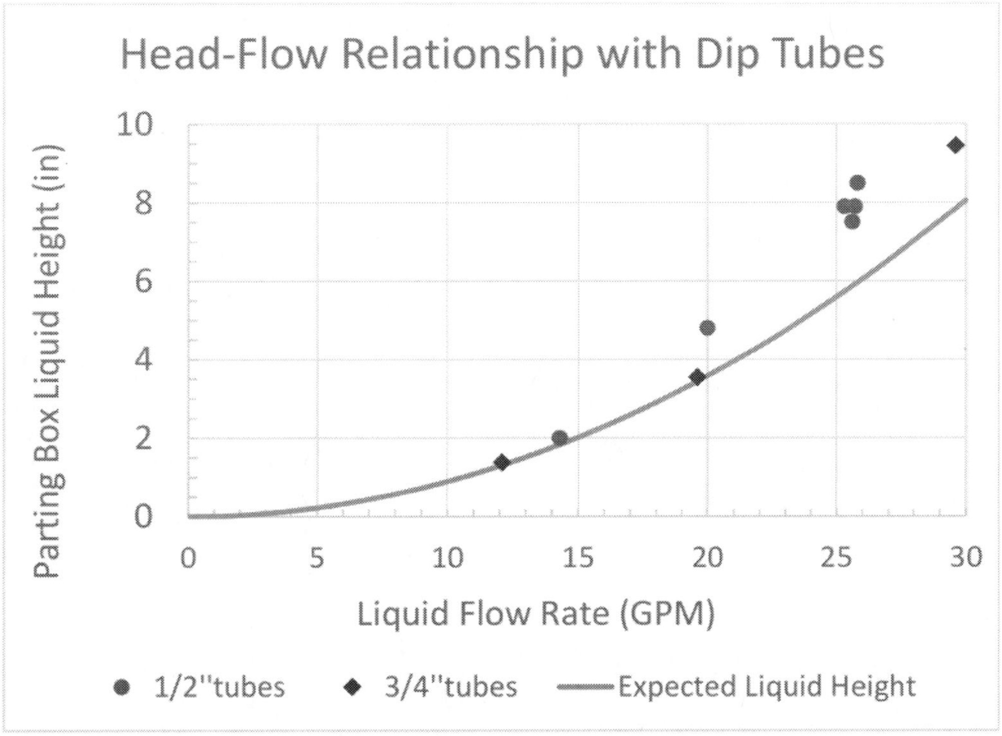

Figure 13. Head- flow relationship with dip tubes. Blue: ½" NPS tubes, terminating 1.75" above floor; purple ¾" NPS tubes, terminating 2⅛" above floor.

Test Results with Dip Tubes with Perforated Tips

Figure 14 shows photographs of the tests with the ½" NPS dip tubes with 40% open area tips. The bottom of the tips was about 5 inches above the parting box floor. Figure 14 shows no aeration at 30 gpm, (when the tips were submerged in the parting box liquid), and some aeration at the lower rates. However, the aeration was visibly subdued as the perforated tips broke the momentum of the incoming liquid jets and dispersed the liquid over a fairly wide area. At the 13 gpm flow rate the sparger pipe was not liquid-full, and some air was sucked into the dip tubes in a manner similar to that described above. The appearance with the 20% open area tip was similar to that discussed for the 40% open area and the relevant photographs are not presented here, but can be obtained from the authors upon request.

Figures 15 and 16 show an excellent split of the liquid to the bins with the dip tubes with the perforated tips, both at the high rates and at turndown to 45% of the maximum. The split was excellent both with the 20% and 40% perforated tips and regardless of whether the water supply line entered perpendicular to the sparger pipe or at 45° to the sparger pipe.

Figure 17 compares the liquid or froth heights as determined visually from the photographs to the liquid heads calculated from the orifice equation (with coefficients per Perry's Handbook[3]). With the 20% and 40% perforated tips, at all liquid flow rates, the measured liquid heads were well-predicted by the orifice equation.

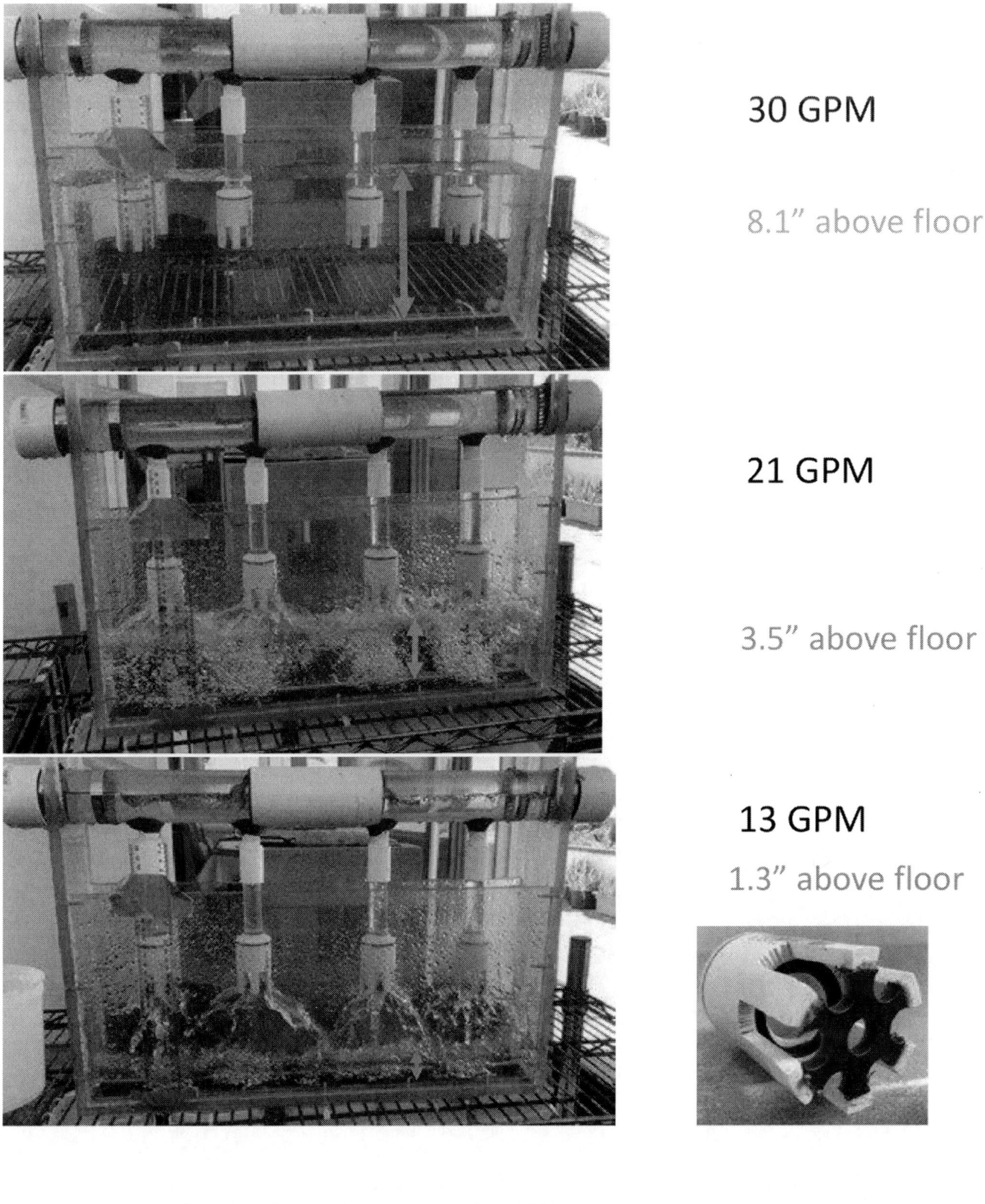

Figure 14. Tests with dip tubes with perforated tips, 40% open area

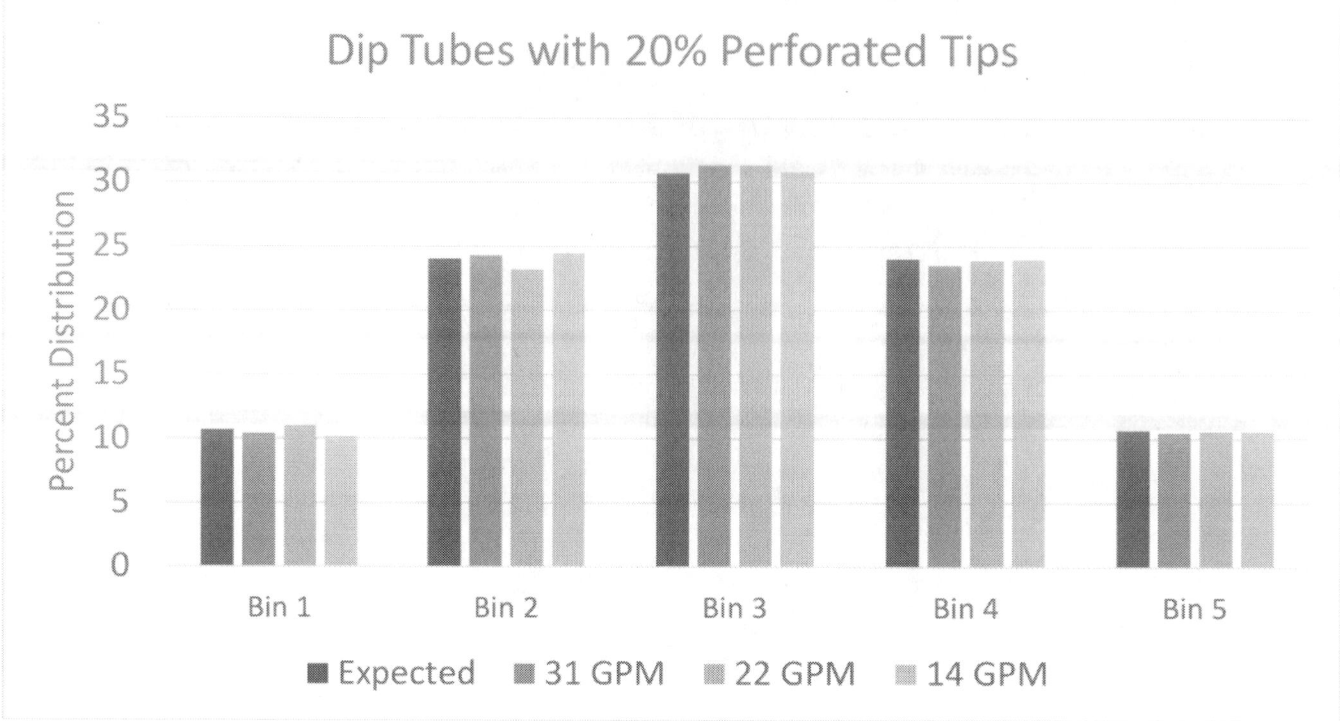

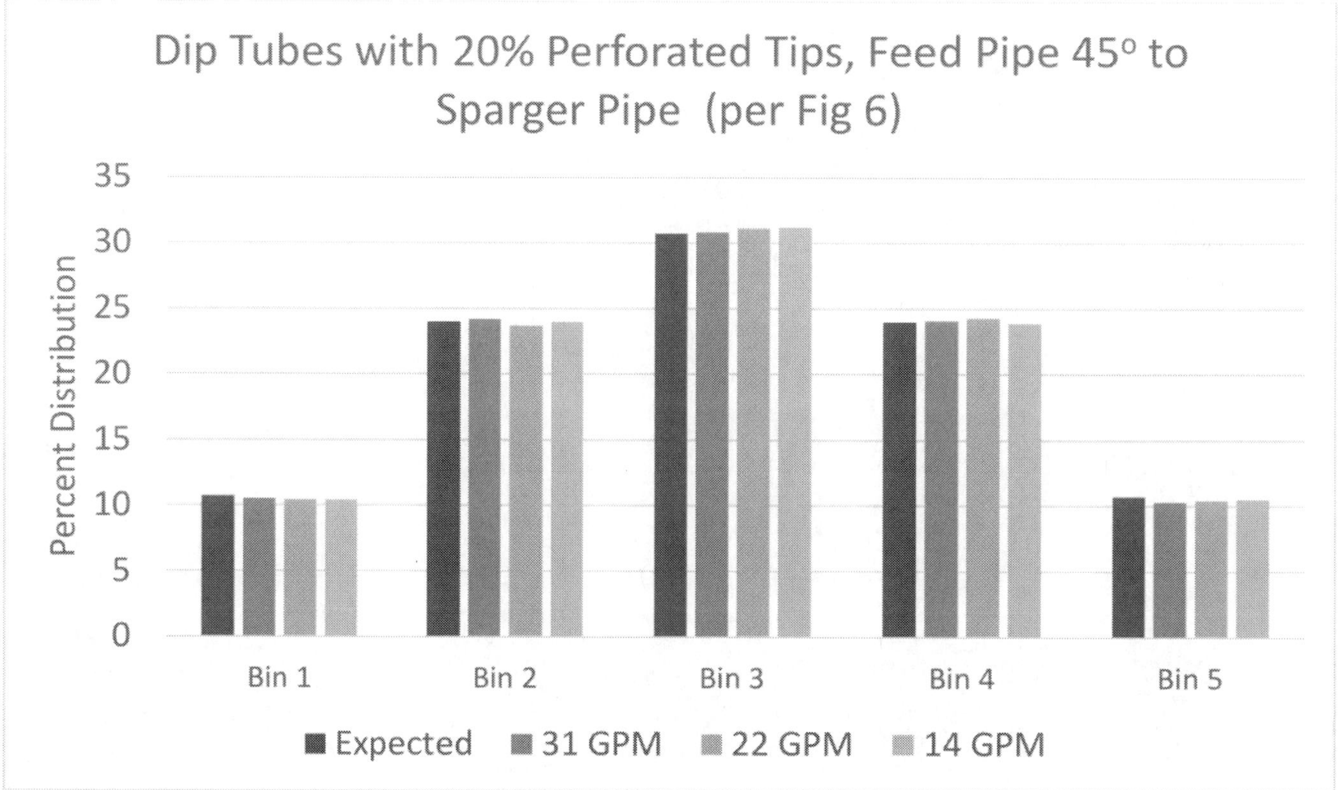

Figure 15. Liquid split among the bins compared to the desired split based of equal liquid flow rate through each perforation. Dip tubes with 20% open area. Upper diagram, water supply line entered perpendicular to the sparger pipe, lower diagram at 45° to it (as in Figure 6).

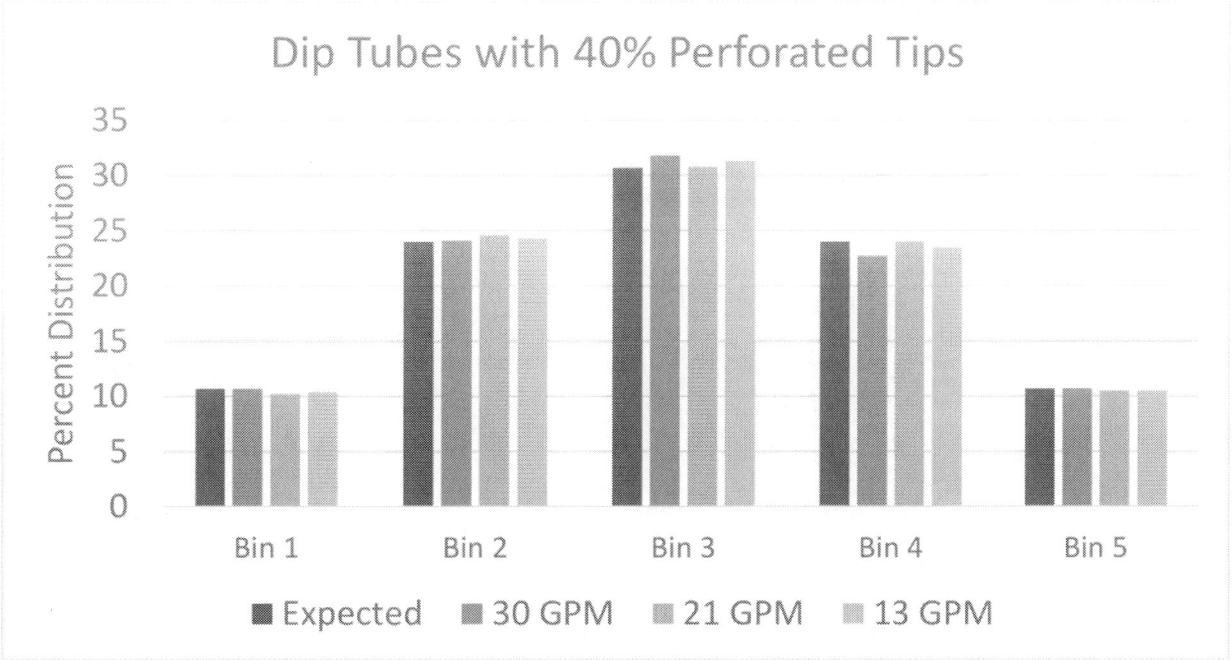

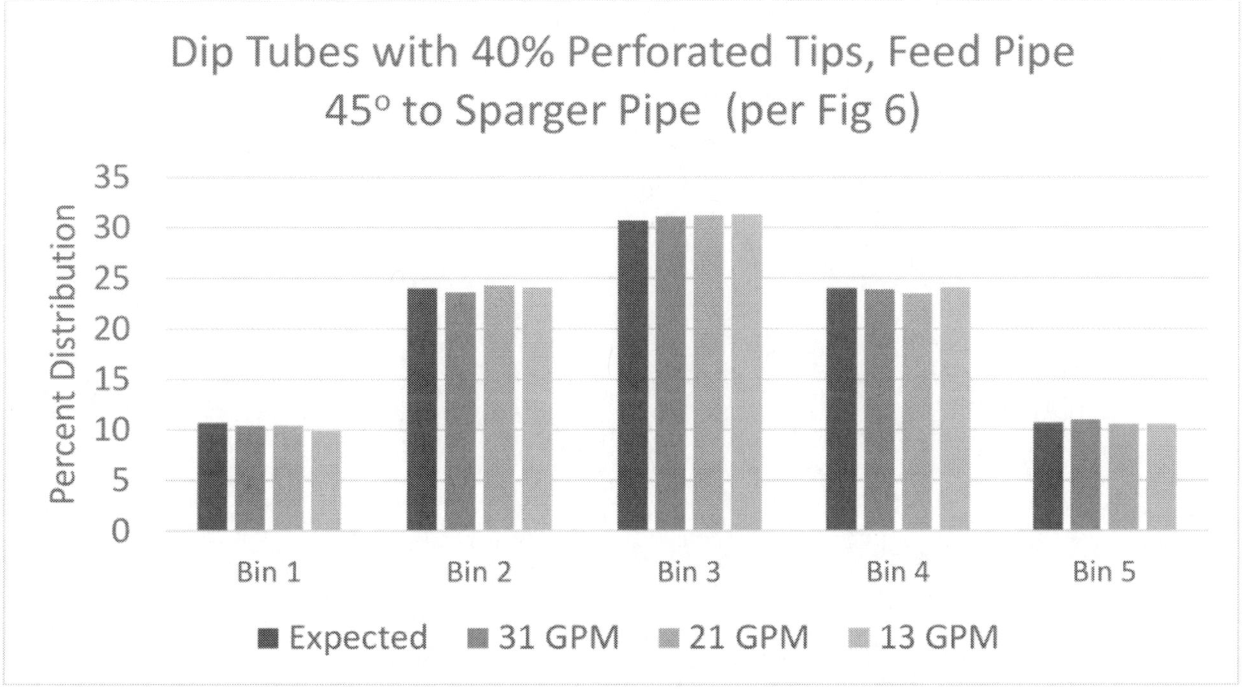

Figure 16. Liquid split among the bins compared to the desired split based of equal liquid flow rate through each perforation. Dip tubes with 40% open area. Upper diagram, water supply line entered perpendicular to the sparger pipe, lower diagram at 45° to it (as in Figure 6).

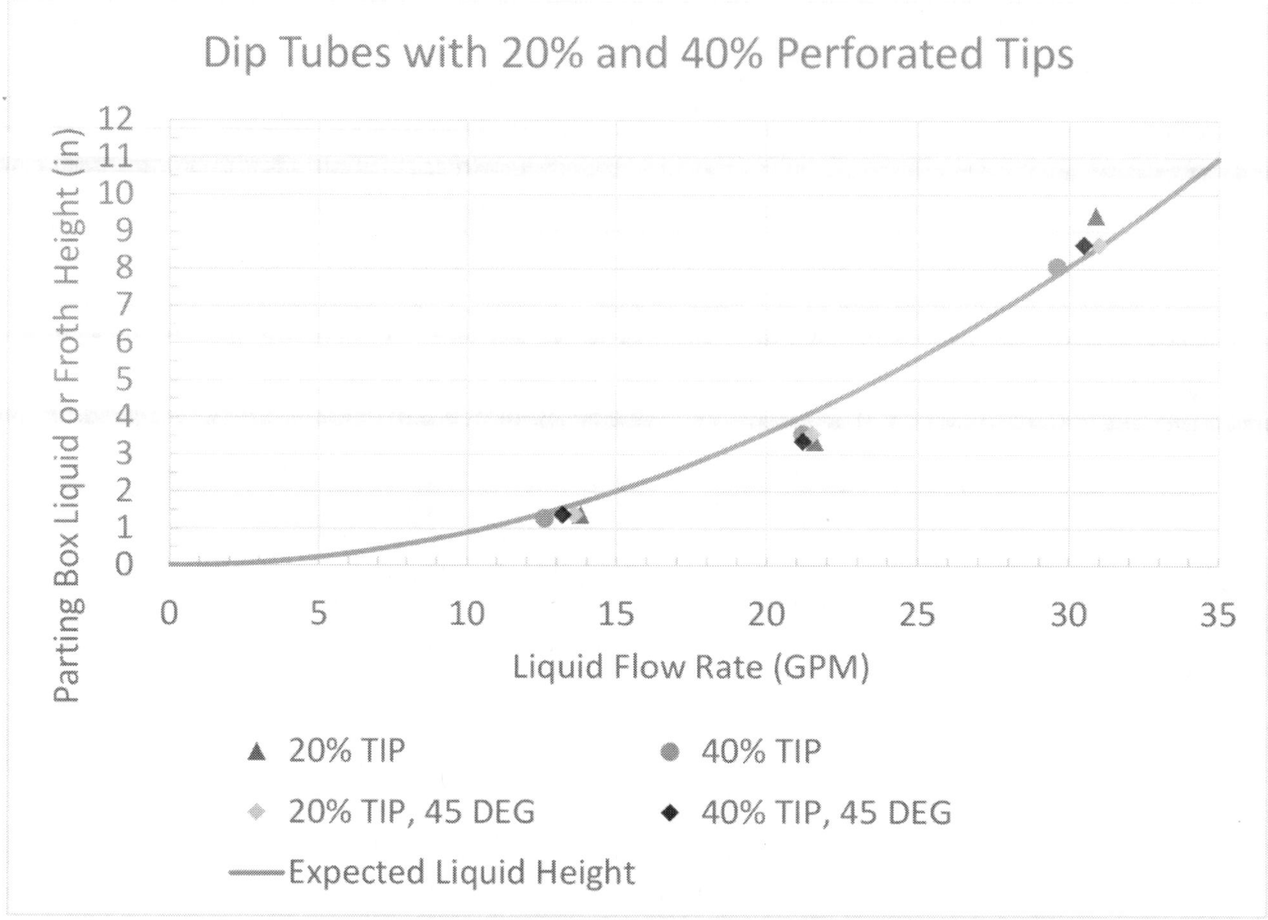

Figure 17. Head- flow relationship, with dip tubes with 20% and 40% perforated tips. 5" above the parting box floor. Water supply line enter perpendicular to the sparger pipe (blue and green symbols), or at 45° to the sparger pipe (as in Figure 6) (yellow and black symbols).

Test Results with Dip Tubes with Spray Nozzles

Figure 18 shows photographs of the tests with the dip tubes terminating in Spraying Systems Co. ¾" K-120 nozzles. Adjacent sprays were directed toward opposing walls of the parting box to break the momentum of the liquid. The bottoms of the spray nozzles were about 4 inches above the parting box floor.

Figure 18 shows no aeration at 30 gpm, when the sprays were submerged in the parting box liquid. However, there was strong wave disturbance observed on the surface of the liquid, as can be seen in the upper photograph of Figure 18. There was intense aeration and turbulence generated by the sprays at the lower rates. At the medium rates (19 gpm), some liquid appears to have been pushed 2-3 inches up the wall by the sprays. At the low liquid rates, the parting box froth appeared more even and less turbulent, but there appeared to be some unevenness in flow between the four dip tubes and sprays. Also, at the lower liquid rates the sparger pipe was not liquid-full, but no air appeared to be sucked into the dip tubes.

Figure 19 shows an excellent split of the liquid to the bins with the dip tubes with the sprays at the high liquid rates, which quickly deteriorated as the liquid rates were reduced. Unlike some of the previous tests in which at turndown more liquid went into the center holes, in these spray tests more liquid went into the off-center holes and less into the center holes.

Figure 20 compares the liquid or froth heights as determined visually from the photographs to the liquid heads calculated from the orifice equation (with coefficients per Perry's Handbook[3]). The measured liquid heads were well-predicted by the orifice equation, except at the maximum rates in which the liquid level in the parting box was about 10-15% higher than predicted. This may be due to the momentum of liquid exiting the sprays pushing the liquid level up in the box.

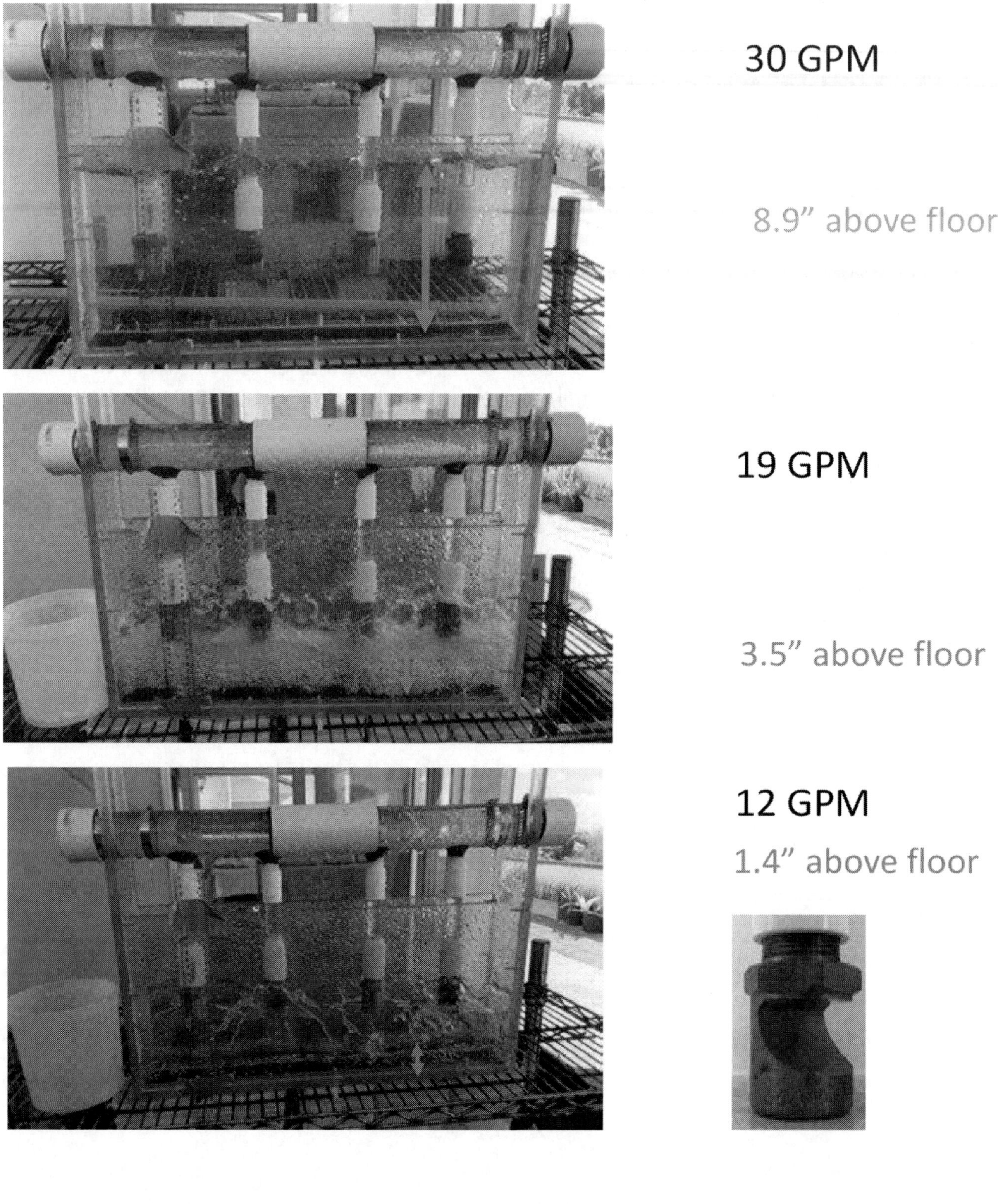

Figure 18. Tests with dip tubes terminating in spray nozzles 4" above the parting box floor, with liquid issuing toward the wall.

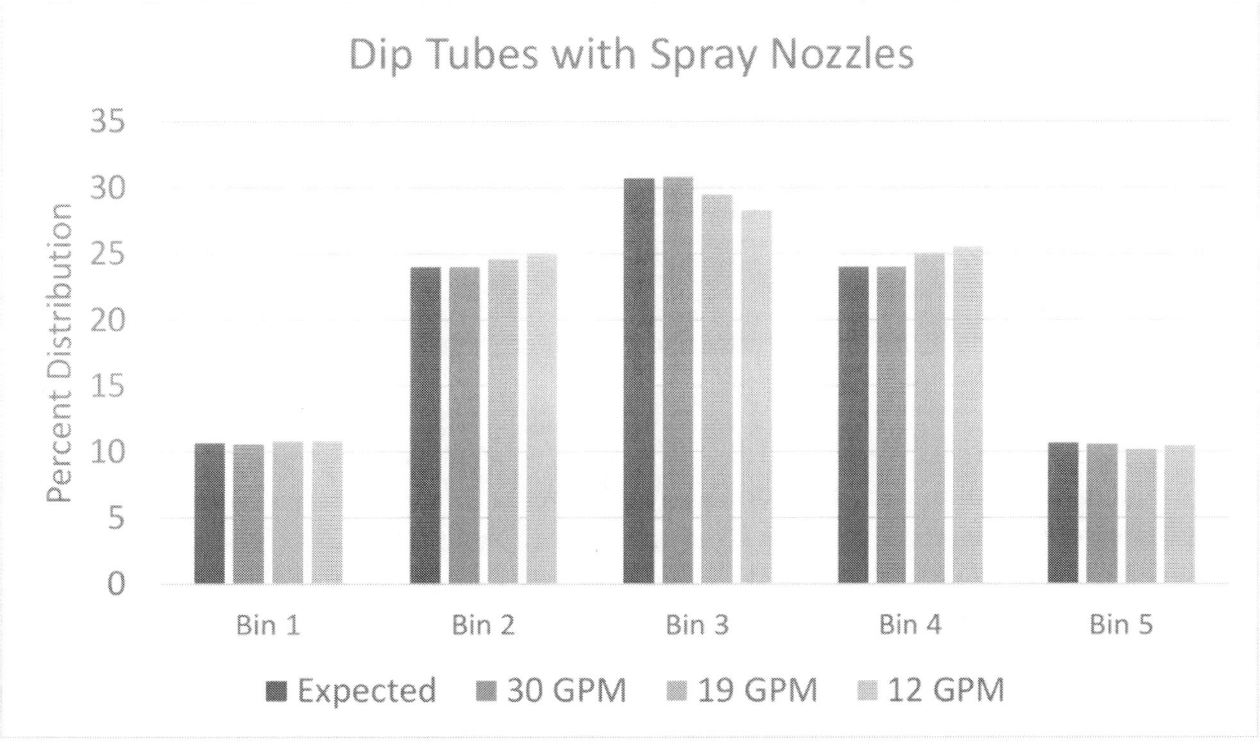

Figure 19. Liquid split among the bins compared to the desired split based of equal liquid flow rate through each perforation. Dip tubes with spray nozzles.

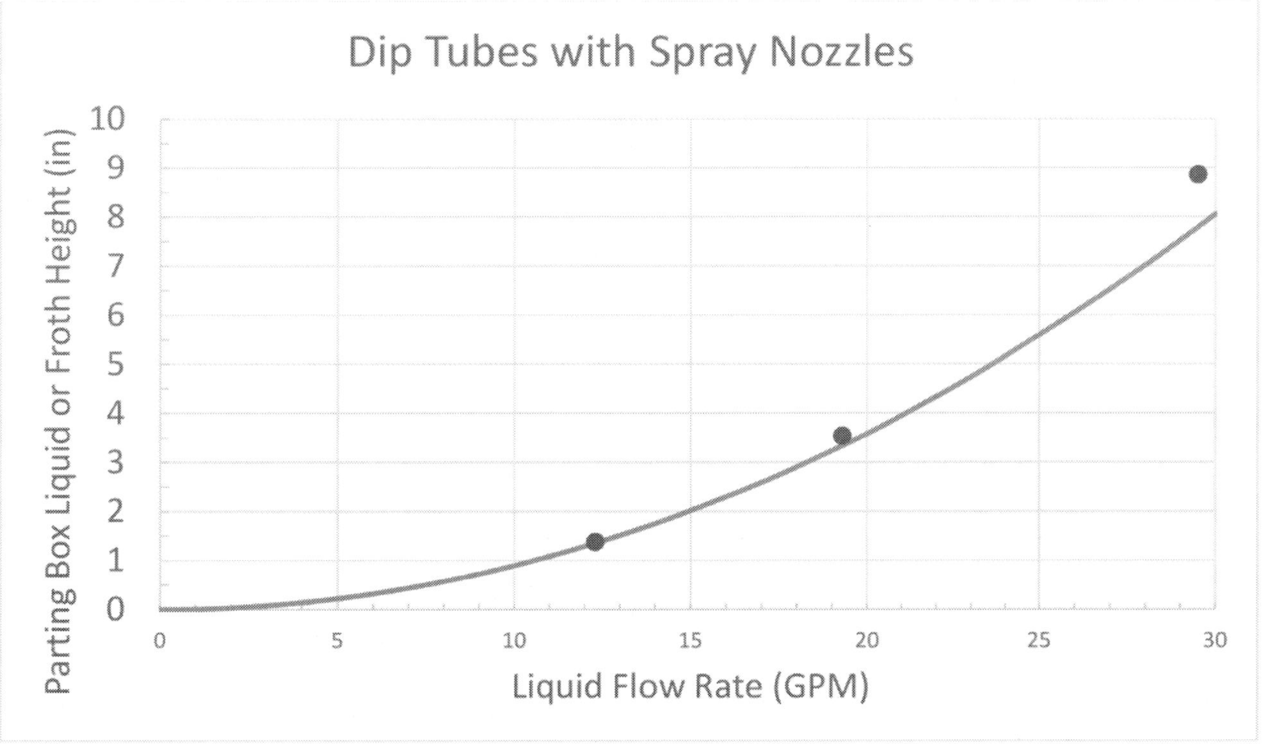

Figure 20. Head- flow relationship, dip tubes with spray nozzles

Orientation of Main Feed Pipe to Feed Sparger

We tested designs that gave good split of liquid into the bins and low deviations from the orifice equation both with the water supply pipe perpendicular to the feed sparger pipe, and with the water supply pipe entering at 45° to the sparger pipe (as in Figure 6). Figures 9, 15, and 16 show that with the 20% VIP and with the tube dips terminating in 20% or 40% tips, the liquid split was excellent regardless of whether the water supply line was perpendicular or at 45° to the sparger. Figures 10 and 17 show that the orientation of the water supply pipe to the sparger pipe had no effect on the head-flow relationship either.

In our previous test series[2] we identified another design that gave good liquid split to the bins and low deviations from the orifice equation: dip tubes terminating with tees. Here we tested this design with the water supply pipe entering at 45° to the sparger pipe (Figure 21). The tees for these tests terminated 6 inches above the parting box floor. Figure 21 shows the high-rate (31 gpm) test.

Figure 22 shows that the excellent liquid split to the bins, previously measured with the dip tubes with the tees[2], remained excellent when the water supply pipe was oriented at 45° to the feed sparger. Figure 23 shows the same for the head-flow relationship. Regardless of whether the water supply pipe entered perpendicular or at 45° to the feed sparger, the head flow relationship with this design was the same and followed the orifice equation throughout the flow test range.

In conclusion, for all the designs that gave good split of liquid into the bins and low deviations from the orifice equation, our tests showed that the 45 degrees bend had no effect on either the liquid split to the bins or on the head-flow relationship.

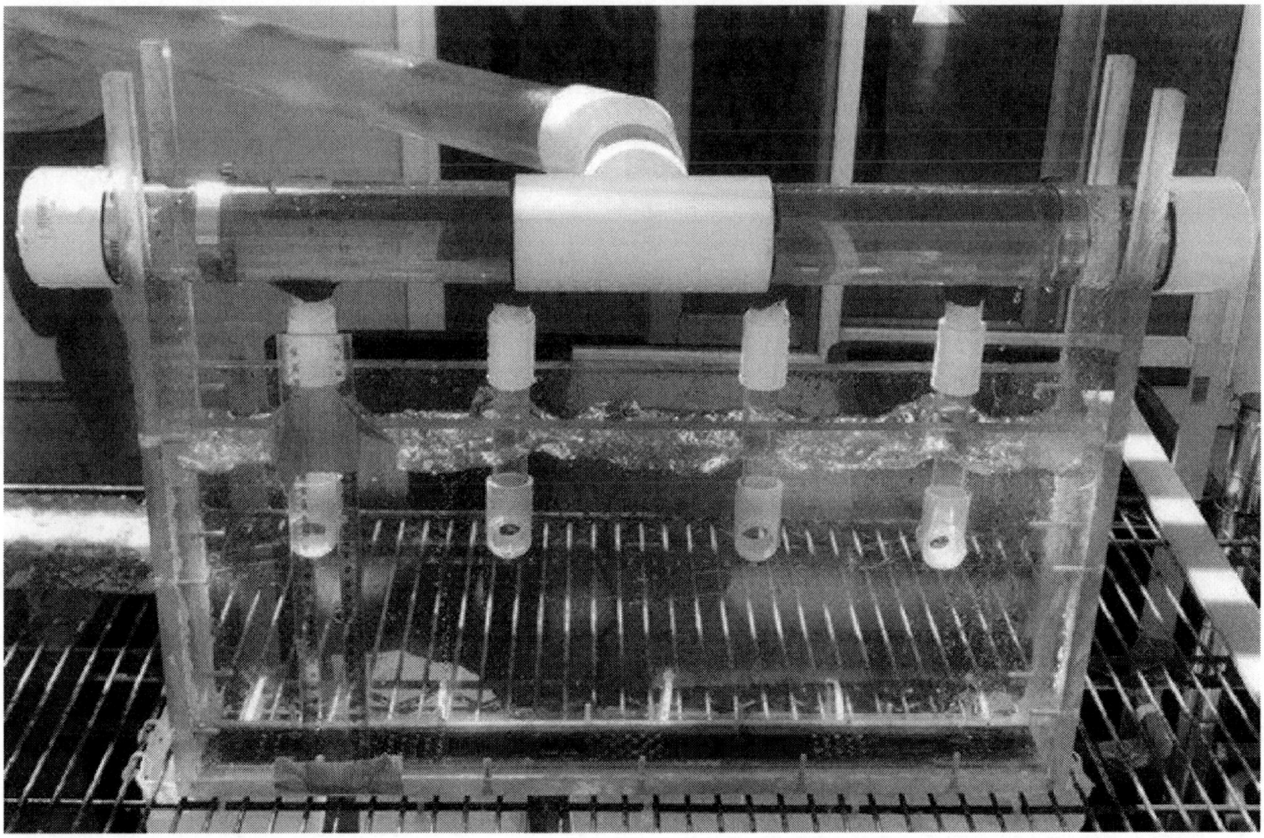

Figure 21. Tests with dip tubes terminating in tees 6" above the parting box floor, with liquid issuing toward the wall. Water supply pipe is oriented 45° to sparger feed pipe. 8.7 inches level, wavy appearance on top.

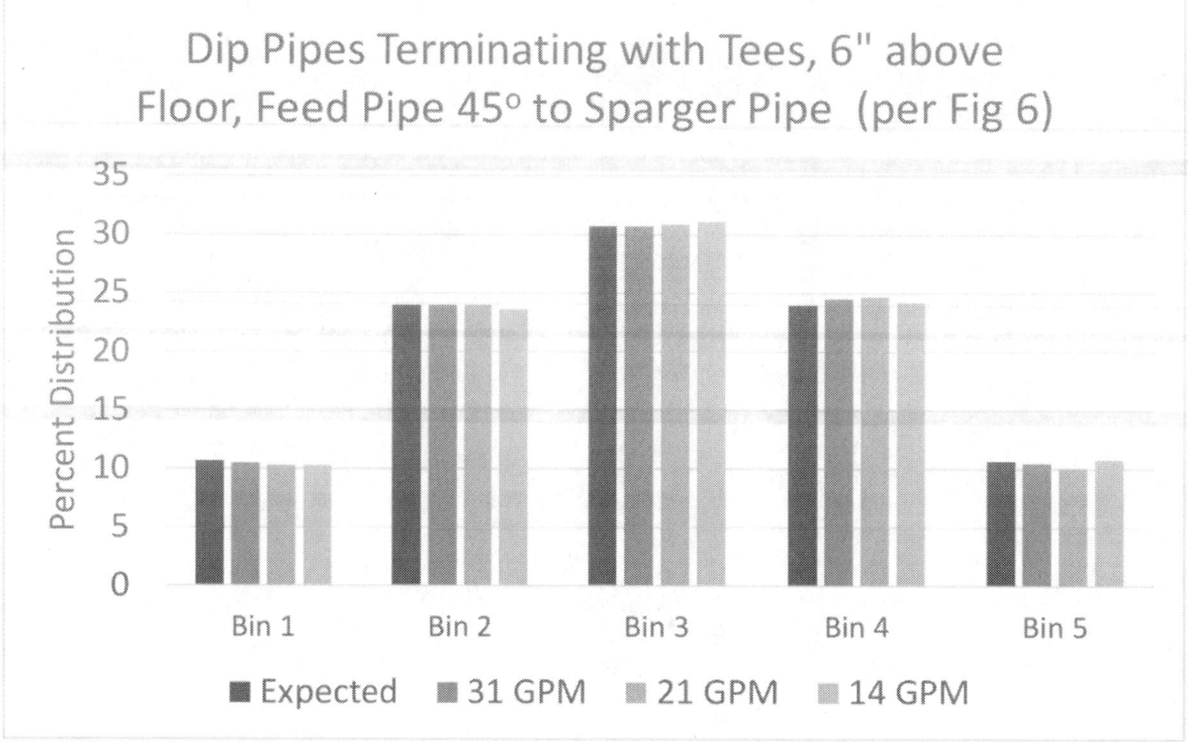

Figure 22. Liquid split among the bins compared to the desired split based of equal liquid flow rate through each perforation. Dip tubes terminating with tees, 6" above parting box floor, water supply line enters at 45° to the sparger pipe (as in Figure 21).

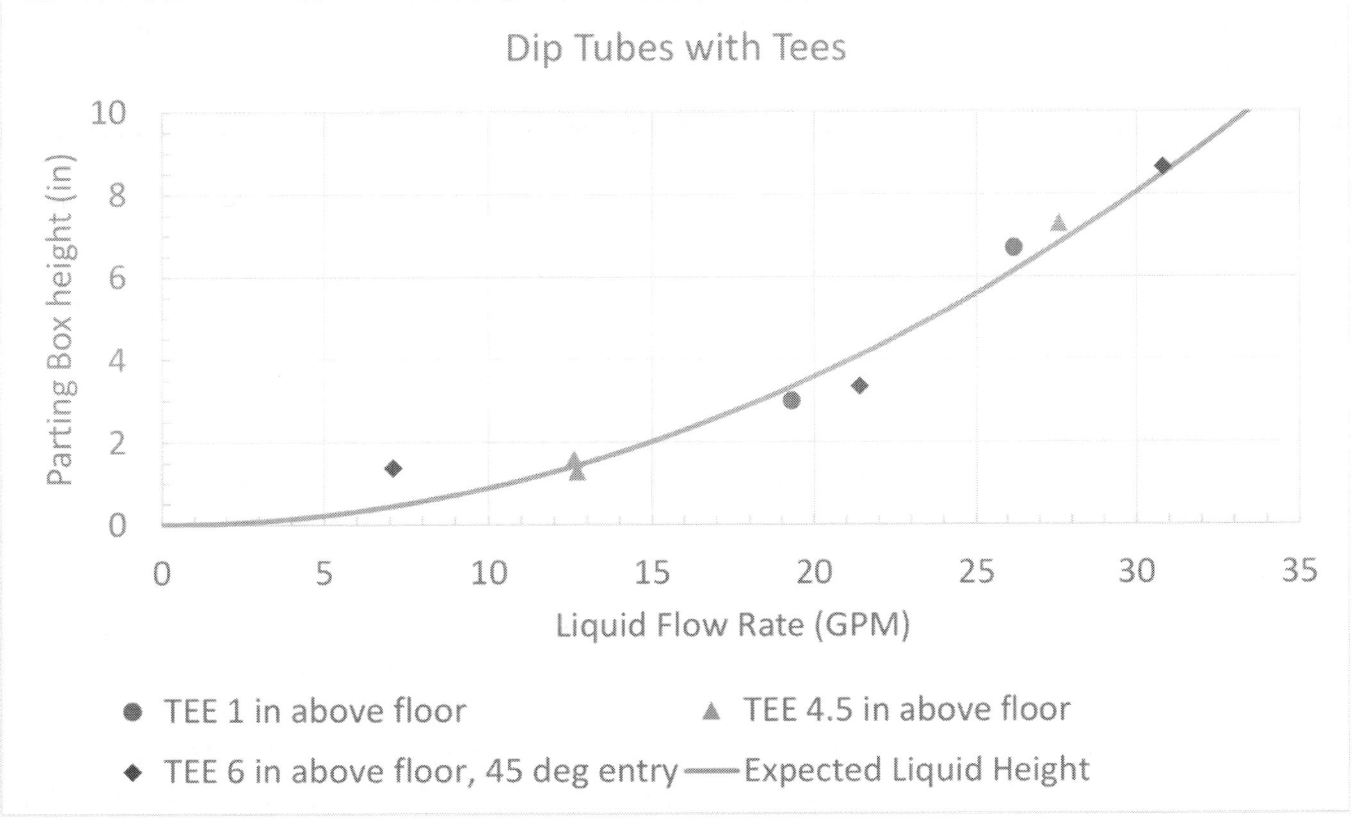

Figure 23. Head- flow relationship, dip tubes with tees, water supply line entering perpendicular to (blue and green) or at 45° to (purple) the sparger pipe (as in Figure 21). Distances above floor refer to the height of the bottom of the tees above the parting box floor.

Conclusions

We performed water tests on a scaled down model of a parting box that exists in a hydrocarbon-separation tower. This model was intended not only to provide insight into the tower issue, but also to provide generic insight into parting box hydraulics as tested in a realistic parting box design. The experiments conducted extended well beyond the specific design under investigation into generic parting box hydraulics.

Our previous tests[1] showed that feeding the parting box from a sparger pipe with downward-pointing holes as recommended by the published literature[4-6] generated extensive frothing in the parting box at both high and low liquid rates. The frothing was due to water jets issuing from the feed sparger pipe entraining air into the water upon impact. This generated excessive froth heights in the parting box, overflowing out of the box at high rates. The liquid split to the five clusters of holes at the floor of the parting box was good at the maximum rates, but quickly deteriorated as the liquid rate was turned down. Later tests[2] showed that reducing the liquid velocity out of the sparger holes by using larger holes reduced the excessive froth height, but the liquid split to the five clusters of holes at the floor of the parting box was still good only at the maximum rates, again quickly deteriorating as the liquid rate was turned down.

Using a Perforated Impingement Plate (PIP) with either 20% or 10% open area, mounted either 2" or 4" above the parting box floor, and extending from wall to wall in the parting box, greatly improved the split of liquid to the clusters of holes at the floor of the parting box. This split was very good to excellent both at the maximum rates and at turndown to about 50% of the maximum rates[1]. However, an air gap formed beneath the PIP, with tall froth above, producing premature overflows at high rates, and low and relatively less aerated froth beneath the PIP. The air gap resulted from the water jets issuing from the feed pipe entraining air with them. This air passed through the PIP and accumulated beneath the PIP, pressuring up the space between the parting box floor and the PIP. This was seen even upon reducing the jet velocity to 2 ft/s by increasing the sparger hole diameter.

In the current tests, we cut the width of the PIP to allow venting between the plate and the parting box wall. The current tests showed that using Vented Impingement Plates (VIP's) with 10 to 20 percent open area, with a vapor rise area of 40-50% of the parting box cross section, gave very good to excellent liquid split to the clusters of holes at the parting box floor. At up to 70% of the maximum rates the measured froth heights were close to the liquid heads calculated from the orifice equation, but near the maximum rates they exceeded those calculated from the orifice equation by 10-20%.

Previous tests[2] looked at a design in which dip tubes bring the incoming liquid 1.75" above the floor of the parting box. This fully eliminated the frothing in all but the lowest rate, but again the liquid heights in the parting box were considerably higher than those calculated from the orifice equation. It may be that the horizontal momentum generated when the liquid jets hit the floor of the parting box were deflected sideways, making it more difficult for the liquid to descend through the parting box holes, causing larger liquid heads. For these tests, the liquid split to the clusters of holes at the parting box floor was mediocre at all the tested liquid rates.

In the current tests, we used larger dip tubes with 2⅛" clearance from the bottom of the parting box to lower the momentum of the liquid. At up to 70% of the maximum liquid rate the liquid height in the box was well-predicted by the orifice equation, but near the maximum rates, the measured liquid heights remained about 15-20% higher than the liquid heads predicted by the orifice equation. For these tests the liquid split to the holes at the parting box floor was excellent at high rates, but deteriorated upon turndown.

Our previous tests[2] showed that fitting the smaller dip tubes with tees that directed the liquid towards the parting box walls eliminated the excessive liquid height problem, gave excellent liquid split to the clusters of holes in the parting box floor all the way to 45% turndown, and a head-flow relationship closely matching the orifice equation. This happened with the bottom of the tees terminating both 1" and 4.5" above the parting box floor.

Our current tests looked at an alternative design with shortened dip tubes fitted with perforated tips, either with 20% or 40% open areas, mounted about 4 inches above the bottom of the parting box. Like with the tees, fitting the dip tubes with these perforated tips eliminated the excessive liquid height problem, gave excellent liquid split to the clusters of holes in the parting box floor all the way to 45% turndown, and a head-flow relationship closely matching the orifice equation.

Additional tests used dip tubes with spray nozzles directed toward the walls of the parting box. This was less successful, as the sprays generated more turbulence in the box. The split of the liquid to the clusters of holes in the parting box floor was excellent at high liquid rates, deteriorating as the liquid rates were reduced. The measured liquid heads were well-predicted by the orifice equation, except at the maximum rates in which the liquid level in the parting box was about 10% higher than predicted.

With designs that performed well, a water supply pipe perpendicular to the feed sparger pipe gave much the same performance as one oriented 45° to the feed sparger pipe.

In summary, it appears that the designs that gave good liquid split to the parting box holes at maximum and turndown rates without inducing excessive froth heights, employed dip tubes with a mechanism to break the momentum of the entering liquid jets, either by using tees directed towards the wall at the bottom or by using a well-designed perforated tip with 20 to 40 percent open area.

Spargers with no dip tubes tend to give excessive frothing and their liquid splits to the clusters of holes at the parting box floor appears to be good only at the maximum liquid rates, quickly deteriorating upon turndown. Installing perforated plates in the lower part of the parting box provided a good split also at the lower liquid rates. Our tests showed that these plates need to be adequately vented to prevent vapor trapping underneath which will induce excessive froth heights, potentially resulting in premature overflow.

References

1. Kister, H. Z., G. Jacobs, and A. A. Kister, "Gaining Insights into Parting Box Hydraulics", Kister Distillation Symposium, Topical Conference Proceedings, p.463, AIChE Spring Meeting, New Orleans, LA, March 31- April 4, 2019.
2. Kister, H. Z., G. Jacobs, and A. A. Kister, "Improved Feed Entry Improves Parting Box Hydraulics", Paper presented at the Distillation Symposium, Topical Conference, AIChE Spring Meeting, Virtual, August 16 - 20, 2020.
3. Kister, H. Z., P. Mathias, D. E. Steinmeyer, W. R. Penney, V. Monical, and J. R. Fair, "Equipment for Distillation, Gas Absorption, Phase Dispersion, and Phase Separation", Section 14, in R. H. Perry and D. Green "Chemical Engineers' Handbook", 9th Ed., McGraw-Hill, N.Y., 2018.
4. Norton Company, "Packed Tower Internals", Bulletin TA-80R, 1974 (available from KochGlitsch, Wichita, Kansas).
5. Fadel, T. M. "Selecting Packed-Column Auxiliaries", Chem. Eng., January 23, p.71, 1984.
6. Kister, H. Z. "Distillation Operation", McGraw-Hill, N.Y., 1990.
7. Senecal, V. E., "Fluid Distribution in Process Equipment", Ind. Eng. Chem. 49(6), p.993, 1957.

618471 Influence of Design Characteristics of Structured Packings on the Foam Formation in Columns

Abstract:

Influence of design characteristics of structured packings on the foam formation in columns

Hannes Leuner, Sören J. Gerke, Jens-Uwe Repke, Technische Universität Berlin, Process Dynamics and Operations Group, Berlin/Germany

Foam in distillation and absorption columns is one of the main reasons for process malfunctions in the chemical industry (Kister, 2003). Disturbing foam in columns usually leads to higher gas pressure loss, lower gas throughput, lower mass transfer rates and therefore a reduced separation efficiency that can make a shutdown necessary (Thiele et al., 2003). Foam is caused by surface-active substances which are either unavoidable or often unwanted, unknown and a result from contamination or conversion of participating substances. Hence, the foam formation in columns is a complex phenomenon with individual causes and impacts for certain equipment and processes depending on various influence parameters like temperature, pressure, fluid properties, design characteristics of the equipment and until now neither completely understood nor fully controllable (Leuner et al., 2020).

To further understand foam formation and develop concepts of design and operation to sufficiently inhibit and destruct foam in structured packing columns, the influence of design characteristics of structured packings on the foam formation in columns has been investigated in this work. Thereby, the influence of the microstructure of the packing material, the contact angle and the junction between packing layers of structured packings was analyzed and discussed with already published findings. The aim of the study is to identify foam inhibiting design characteristics of structured packings to derive recommendations for future design and operation of foam affected distillation and absorption processes.

The experimental investigations were conducted with Sulzer Mellapak 350.Y in either a DN300 glass column with a packing height of 2m (see Figure 1 left) or a customized measurement cell made from transparent moldings of the structured packing sheets (see Figure 1 right). The foam formation has been qualitatively and quantitively analyzed using optical measurement methods like Light Induced Fluorescence in combination with a high-speed camera along with the gained datasets of gas pressure loss and gas flow rate across the structured packing. Within this study the foam formation of Butanol/Water- and Monoethanolamine/Water-mixtures have been investigated.

Current results indicate a lower foam impact on low liquid loads in

Figure 1: Experimental set-ups to investigate foam formation in structured packings (1) DN300-glass columns with structured packing. (2) Measurement cell with transparent moldings.

structured packings with smooth surfaces due to a decreased liquid accumulation at contact points between packings sheets. For higher liquid loads, structured packings with a corrugated surface tend to be less affected by foam formation due to an enhanced liquid distribution across the structured packing. However, hydrophobic surfaces with contact angles above 90° are in all cases more affected by foam formation than hydrophilic surfaces with contact angles below 90°.

Kister, H. Z. (2003). What caused tower malfunctions in the last 50 years? Chemical Engineering Research and Design, 81(1), 5-26.

Leuner, H., Gerstenberg, C., Lechner, K., McHardy, C., Rauh, C., & Repke, J. U. (2020). Overcoming unwanted foam in industrial processes of the chemical and food industry an ongoing survey. Chemical Engineering Research and Design, 163, 281-294.

Thiele, R., Brettschneider, O., Repke, J. U., Thielert, H., & Wozny, G. (2003). Experimental investigations of foaming in a packed tower for sour water stripping. Industrial & engineering chemistry research, 42(7), 1426-1432.

... *a project of the* ***Industrial Collective Research (IGF)***

Supported by:

Federal Ministry for Economic Affairs and Energy

on the basis of a decision by the German Bundestag

AiF Forschungsnetzwerk Mittelstand

FEI FORSCHUNGSKREIS DER ERNÄHRUNGSINDUSTRIE E.V.

iGF Industrielle Gemeinschaftsforschung

This IGF project of the FEI is/was supported via AiF within the programme for promoting the Industrial Collective Research (IGF) of the German Ministry of Economics and Energy (BMWi), bases on a resolution of the German Parliament.

618759 A Real-World Case Study: (Re)Assessing Distillation Tower Instrumentation & Controls for Optimal Operations

Abstract:

Distillation towers constitute a significant portion of unit operations in every Hydrocarbon Processing Industry (HPI) facility. While the thermodynamic principles and mechanical design of these vital pieces of equipment are well understood, often times it is the design & maintenance of the instrumentation and controls that become the rate-limiting factor affecting their on-going reliability and real-time performance.

A specialty chemicals facility was exploring how to minimize product yield losses while maintaining quality in several of its existing distillation trains. The original instrumentation & control schemes had become less reliable and generated unsatisfactory results as the units were pushed for greater production and recovery, requiring manual intervention & control by the Operators. A process control consultant was asked to assess the existing controls structures & instrumentation and recommend revisions that would achieve the desired objectives without requiring operator-intervention. The Consultant's activities centered around answering the following questions:

- What is the appropriate measurement that best infers the quality objective(s)?
- Are there any instrumentation or equipment deficiencies that must be resolved?
- How should the controls be structured to reliably achieve the plant's objectives in real time?

This paper describes the analysis procedures and methodologies that were used to generate the recommended instrumentation & control modifications, as well as some of the surprising (and not so surprising) discoveries that were made during the assessments. The preliminary results of the recommended control modifications will also be presented.

SPRING21
+17TH GCPS
A Joint AIChE and CCPS Meeting

Tool for Safety & Organizational Control: SOC - Just Culture

Salvador Ávila
UFBA
avilasalva@gmail.com

Gunther Josuá Costa
UNIVASF
guntherjosua@gmail.com

Luiz Fernando Lopes de Souza
COPENOR
lflsouza@globo.com

Lucas Pereira Menezes
UFBA
lucasmpereira13@hotmail.com

Prepared for Presentation at
American Institute of Chemical Engineers
2021 Spring Meeting and 17th Global Congress on Process Safety
Virtual
April 18 - 22, 2021

Tool for Safety & Organizational Control: SOC - Just Culture

Salvador Ávila
UFBA
avilasalva@gmail.com

Gunther Josua Costa
UNIVASF
guntherjosua@gmail.com

Luiz Fernando Lopes de Souza
COPENOR
lflsouza@globo.com

Lucas Pereira Menezes
UFBA
lucasmpereira13@hotmail.com

Abstract

This work was designed to present to NETWORK Research and Services, a demonstration of the concepts and applications of real cases in Human Factors, Risk, Reliability and Crisis. Although we work with Complex Systems that are difficult to predict (such as the chemical process industries) and understanding the nature of the processes in Sociotechnical Systems, we can build Rules and Barriers that have better results than current ones. Concepts are the basis, tests are confirmation, algorithms indicate associated methods and techniques, and validation indicates the certainty of achieving good results. In the case of Reliability, the concepts and algorithms were tested with real cases from the Oil Industry.

The new concepts are systemic failure; cultural force field; socioeconomic and affective utility of the task versus feedback; causal link of low-risk perception; difference between enabling elements of hazard energy versus human factors in the operational situation; management models to avoid accidents and to avoid energy losses and respective techniques that include the discussion about culture and human factors.

A discussion of cultural threats of behavior change (example: blame culture and generational conflicts) and application cases indicating solutions to bad habits and incorrect decisions, they make it easier to understand why social risks are dynamic and do not separate from the routine.

The human elements are part of the project and cut (disable) the how of dangerous energy. The intention is that these elements of the safety culture and the technological project are safeguards, but it may be that the project is incomplete or that the barriers have suffered social and physical threats, reducing resistance in relation to the load. These elements of the managerial (stress and leadership), technological (risk and complexity) and behavior (4Cs) dimensions will be limiting the passage of energy by human factors (which includes management, culture, organization, and technology). The 9 levels of human factors that will compose the Bayesian network are: Culture, Social Phenomenon, Company, Management & STAFF, Group and Individual, Operational Control, Failure Control, Accident and Crisis or Disaster.

The Investigation on the Operational and Technical Culture indicates the formal and informal rules of operation of the team in the Operational Routine, allowing the prediction of the future behavior of the worker, the process, and the equipment from the signs of failure. It is possible to calculate the level of approach of the discomfort zone for safety events from this plant history. For this we use, data mining, grouping, analysis of main components, definition and calculation of the objective function, heuristic rules, and adjustment of factors through fuzzy mechanics.
This solution-guiding document addresses events with different stress levels in which the operating team lives: operational deviations, uncontrolled processes, damage on equipment, and task failures, safety events such as fire, and catastrophic events that unfold for crisis. Theoretical models are presented, discussion of human error in emergencies, and proposal of appropriate competence for emergency brigade.
The internal and external regulation after these discussions needs to revise its principles to correct the standardization and organizational communication processes, which would change the procedures for team building and the criteria for the technology project. Thus, the discussion on the regulation and formatting of human elements and human factors will clarify technicians and managers as to what should be done. An improvement program that includes researching lessons learned is included in the work.
Finally, here technological solutions are suggested in the technological, managerial, social and organizational dimension for the best control of companies. Products and Processes, Courses, Interventions in Human Factors to review concepts and habits. Based on the experience of a project in the metallurgical industry, we propose interventions to adjust the safety culture. Also, for better guidance on mathematical and statistical techniques, we present a roadmap and the reference area. A brief presentation of the techniques for identifying and measuring behavior variables (enabling elements) is carried out.

Keywords: reliability, human factor, organizational culture, just culture, safety

1 Introduction

A culture is only configured, as one of the highest levels of maturity in the safety and organization culture, which must be present in the various industrial and / or service domains, where risk management in industrial domains is exceptionally reliable (HRO). It is important to establish high reliability to avoid or mitigate the consequences of accidents for human life, the environment or even property losses. With this, maintain the positive image of the company and the good operational and financial performance of its activities.
The first safety culture studies emerged from the nuclear disaster in Chernobyl [1], where the contribution of a set of organizational factors was evidenced, impacting human factors in the operations related to the accident. Since then, studies in the field of safety culture have gained increasing attention not only in the scientific field, but also in the practical field of industrial activity.
Reason [2] was the one who first addressed the so-called organizational issues or latent factors to be considered in the development of barriers to the spread of dangerous energy that could lead to an event of a catastrophic nature. However, we are around a series of social transformations, of a dynamic nature and often - combinatorial and difficult to predict -, which enter the industrial facilities, interfering in the work teams, in their relationship with their supervisors, in the level of perception risk or even in the generation of a culture of risk aversion.

Nowadays, due to the social dynamism in which we live, industrial organizations are no longer a set of systems of a technical-technological nature, to increase the complexity of the organizational environment with the social construct, resulting in new relationships and impacts. Thus, risk management increasingly assumes a dynamic character: time influencing, time being influenced.
In this sense, it is necessary to discuss new concepts in the socio-technical context from the technical or operational culture of the organization, before seeking tools for decision making. These new concepts concern: systemic failure and dynamic risk [3]; cultural force field [4]; socioeconomic and affective utility of the task versus feedback; causal nexus of low-risk perception [5]; difference between enabling elements of hazard energy versus human factors in the operational situation; management models to avoid accidents and to avoid energy losses and respective techniques that include the discussion on culture and human factors.
The characteristic failures of the socio-technical system, understood as systemic failure, are the result of behavioral heuristics acquired and indicated from human errors, incorrect decisions and bad habits related to processes, equipment, and products. Therefore, the operational culture is related to the task and the informal rules assumed in the operation.
The Investigation on Operational Culture indicates the formal and informal rules of operation of the team in the Operational Routine, making it possible to predict the future behavior of the worker, the process, and the equipment from the signs of failure. It is possible to calculate the level of approximation of the discomfort zone for safety events from this plant history [6]. For this we use, data mining, grouping, analysis of main components, definition and calculation of the objective function, heuristic rules, and adjustment of factors through fuzzy mechanics.
The event that generates an operational culture that differs from organizational and safety culture, moving away from the Just Culture model [4], where leaders are actively listening, can assume a chain reaction dimension with high energy release of danger causing scenarios of loss of production or property, or even culminating in disaster. In this way, the process safety dimension is surpassed, where the actions are internal despite the impacts being, even, external. Public Security calls on state and neighboring Security agencies to act and deal with contingencies. The discussion starts to include external barriers to avoid the release of great energies in crisis scenarios.
In fact, such a context posed a new challenge for organizational leaders who try to control organizational human performance in a socioeconomic and dynamic environment. The basis of this context and of the dynamic perception of the interaction of its constituent elements are the result of research work resulting from pilots in the industrial and service areas during the last 20 years, in industrial sectors such as petrochemical, fertilizers and refining. In this sense, all works had real application in industry, services and even in regulatory bodies [10,11,13,14,15]. These works are cited and referenced to direct more knowledge before real applications.
Thus, the organizational leadership starts to play a predominant role in perceiving and dealing with these dynamic and difficult to predict aspects present in its dynamic operational culture and adjusting the inherent dimensions of the organization's safety culture. Highlighting the values of safety and the importance of risk perception, as well as the means or ways that the organization transmits communication and coordinates its actions and decision-making between all organizational levels and improves its learning mechanisms; it is without a doubt, the great challenge of the organizational leaders for the construction of a culture called Just.
The initial starting point is to discuss cultural threats of behavior change (example: Blame Culture and generational conflicts) and application cases indicating solutions to bad habits and incorrect decisions, which makes it easier to understand why social risks be dynamic and not separate from

the operational routine. Thus, the Operator's Speech analysis [6, 13] has a preponderant role and constitutes an essential “locus” for research and analysis that promotes safer behavior and a operational culture more aligned with safety & organizational culture: Just Culture.
This work contributes to the proposition of a new set of solutions that systematically integrates the migration from an existing context of an operational Blame Culture to a more harmonious context of trust, reciprocity of communication and continuous learning; that is, directing organizational maturity towards a “Juster” culture. To illustrate and highlight this proposition, a framework was developed where concepts, methods and tools are presented in each of its layers that meet the integration and systematization of the new concepts highlighted.
The challenge of this framework is to "close the gap" between the shop floor and top management and their operational and safety strategies and to interconnect new concepts and contexts of the organizational and safety culture with the task. And the key to this is the analysis of the operator's discourse, aiming to reorganize the fragments of information in the workplace combined with new dimensional forces that influence the organization's operational culture; since, in the end, the purpose is to mitigate failures, incidents, accidents that can result in loss of quality in products and processes, impacts on the environment and, even in property losses and in human lives. The presentation and discussion of the referred framework will be resumed, in more detail, in items 3 and 4 of this research.

2 Fundamentals

2.1. Sociotechnical Systems: Complexity and Dynamic Risks

The operational culture is the starting point when trying to understand its deviations, failures, incidents and accidents; but, with the growing and increasingly complex challenge of providing interventions for improvements or gains of an integrated nature in industrial and service organizations (private and public). This integration concerns the set of dimensions of organizational performance: efficiency of processes, resulting in gains in productivity and quality; the organization's financial goals that need to be achieved and maintained; external regulation to be met; development and management of Process Safety and Management Systems and Health, Environment and Occupational Safety (HSE); among others, consistent with the industrial segment in which they operate.
Thus, new concepts and approaches need to be discussed and considered in this challenging context for organizational leaders and, also, leaders in the public sector. Next, we will discuss a body of linked knowledge and the state of practice in HRO domains, which will assist us in identifying potential research and intervention gaps for improving organizational performance (including safety performance). from the theater (contexts and dimensions) of operations and tasks on the shop floor - the operational culture.
The security tools and methods aimed at classic technical systems are already effectively out of step with the current and imminent future context. In the 1980s, Perrrow [7] already addressed the notion of increasing the complexity of technical systems making them socio technical. And that, this was due to “new characteristics of contemporaneity” with a high effect of interdependence, object of study of the Theory of Complexity and contextualized for a greater range of systems domains.
Thus, sociotechnical systems, due to their complexity, are configured as a multidimensional construct that brings challenges to organizational leaders, trying continuously to investigate their interactions and intervene in a more appropriate or approximate way in the context of their

industrial practices and service [8]. These complexities, which can be categorized (under the scope of management guidelines) as: (i) many dynamically interacting elements; (ii) wide diversity of elements; (iii) unanticipated variability and (iv) resilience [9].
The inherent complexity of sociotechnical systems implies recognizing that the risks are of a dynamic nature and this poses great challenges to the management of the dimensions and variables of its operational cultures. The **characteristic failure of the sociotechnical system, known as systemic failure** [10][11], is the result of behavioral heuristics acquired and indicated from human errors, incorrect decisions and bad habits related to processes, equipment and products. In this sense, the operational culture is related to the task and the informal rules assumed in the operation. Knowledge about systemic failure allows integrating the reliability calculation based on factors that connect the different functions (maintenance, operation, process, people, culture and leadership). This systemic failure is built because of structural design deficiencies or situational deficiencies during factory operation. These deficiencies may be the result of inadequate cultural characteristics that affect the safety culture, but they may also occur due to individual triggers of failure arising from specific points in the production system, thus resulting from feedback anomalies and relationships with factor-elements humans. To reduce the probability of failure and loss of energy, it presents the management models for the dynamic risk resulting from human, social and climate change factors.
In the Models and Algorithms for Systemic Failure [6][12], the theoretical models and the discussion on the integration of reliability of different natures are worked on: (i) Social and Security Culture to understand the result of the cultural magnetic field in operational culture to adjust the Safety Culture; (ii) Investigation on the Relationship between Human Factors-Elements and the consequence in the Perception of Risk. These relationships depend on feedback, are intricate, complex and have elaborations on the failure energy. As a result, the normalization of the deviation; and (iii) Management Models in the Project phase to meet the utility of the business in society in terms of avoiding accidents and energy losses, business and compliance risk analysis, technology and organizational design, and operational control.
The integration of the Reliability dimensions takes place through the Systemic Failure Model or Fingertip Model [6][12] and the Failure Model in Sociotechnical Systems. These models were developed to indicate the life cycle of the fault, the latent and top event regions and where to best install the barriers, according to Figure 1.
The focus on the analysis of the issues and the influence of the inherent factors of the operational culture is of paramount importance for the calibration of the organizational and safety culture. The characteristic failure of the sociotechnical system - systemic failure - is the result of behavioral heuristics acquired and indicated from human errors, incorrect decisions and bad habits related to processes, equipment, and products. Thus, the operational culture is related to the task and the informal rules assumed in the operation.
The human elements, which are part of the project, are classified as organizational culture, security, and technology. These elements can degrade during the time, from the project, going through different situations during the operational action indicating that the energy of the events goes or not towards the accident, depending on the operational culture installed on the factory floor. The event that generates the operational culture, unlike the organizational and safety culture, moving away from the Just culture model, where leaders are actively listening, can assume a dimension of chain reaction with a high release of danger energy causing disaster scenarios. In this way, it goes beyond the process safety dimension, in which the actions are internal despite the impacts being external. Public security calls on state and neighboring security agencies to act and deal with

contingencies. The discussion starts to include external barriers to avoid the release of great energies in crisis scenarios.

Human factors are part of the process structures in operation. Some events are strong, having enough "dangerous energy" to pass through the current human elements after the design phase. Some structures (social or technical) can degrade and respond differently to each real operating situation, giving unexpected behavior.

Here it stands out that, there are human factors interconnected with each other that can create risky energy, compared to projected barriers, human elements, this energy can pass or not, increased or in the same proportion. If during operation some structures are at fault, when combined, they can increase "dangerous energy". We started to study what are the levels of human factors that go through the same human elements, whether the accident is possible. Human factors, therefore, are divided into 9 levels that are controlled by human elements, enabling, or disabling (HE) the passage of the dangerous energy flow. The dimensions discussed by Ávila are [14]: culture, management, operation, and failure. The levels of human factors are: (1) HFL1 - Culture and Politics; (2) HFL1 - Social Phenomenon; (3) HFL3 - Company or Organization; (4) HFL4 - Management and STAFF; (5) HFL5 - Group and Individual; (6) HFL6 - Operational Control; (7) HFL7 - Failure Control; (8) HFL8 - Accident Control; (9) HFL9 - Disaster Control in 4.2.

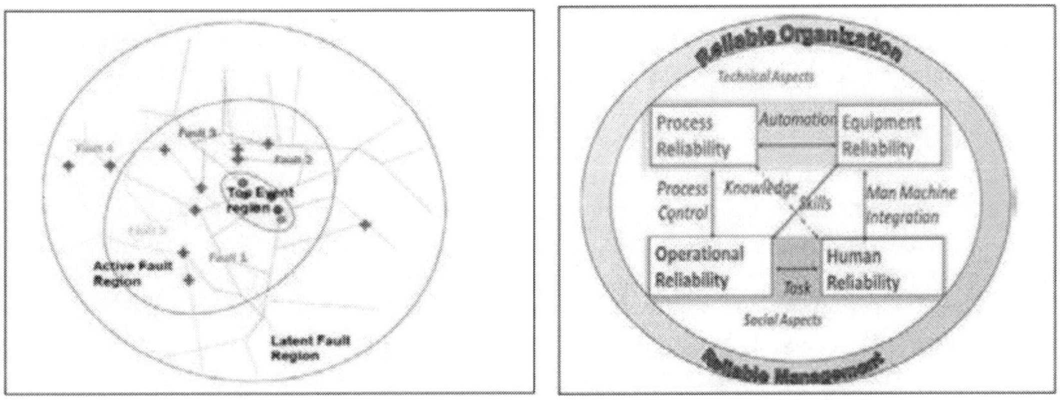

Figure 1 – Sociotechnical Reliability Models [10, 11, 12]

2.2. Operational Culture: Task & Heuristics

Operational Culture is the result of a cultural "magnetic" field, receiving influences from social culture: regional and global, considering the individual and educational characteristics of the organizational participants [4]. Some distortions of the Operational Culture are combated with the organizational and safety culture, others are not corrected. Thus, the operational culture responds to sets of influences, partly static and partly dynamic.

Organizational procedures, rules, criteria, policies, standards of behavior, rites, rituals, expectations, control tools, information flow management, types of communication, training and selection of people are planned, executed, monitored, and controlled, resulting in ways the worker works. The influence of cultures, psychological profiles and educational models can alter the behavior resulting from the design intent, where the measurement of behavior is compared with the standards expected during the performance of critical tasks. Cognitive-affective expectations may not be met considering that operational risk is dynamic. Thus, the response in the use of memory, in the perception of risk and in the accomplishment of the task does not reach the

established goals. The diagnosis on the operational culture indicates the need for adjustments in the organization to meet the goals of safety and integrated management.
Cultural hazards promote an environment conducive to the low quality of human elements that enable hazardous energy based on design barriers to failure and accident. People who live in a cultural environment absorb regional rules or customs, and if they are inserted in mass consumption, they absorb the rules of global culture. Thus, addictions or bad habits can cause human errors or incorrect decisions in a technical environment.
The risks of regional culture, based on Brazilian culture, indicate underreporting because of a culture of guilt that imposes a dynamic of three participants: government, businessman and worker [15;16]. The behavioral characteristics resulting from biases in the regional culture are risk aversion, opportunism, excess flexibility, paternalism, and viewer posture. The behavioral characteristics resulting from global culture bias are generational, gender and / or conflict arising from the multicultural environment. Another problem is the excessive use of the internet bringing virtual impressions in the real world, the impacts can occur on the workstation.
The investigation to detect the culture biases that provoke bad habits and that on the other hand can cause human error or the wrong decision depends on the analysis of archetypes to identify repetitive errors and recirculation of the danger energy in addition to the analysis of the executive function. Published works: (a) Blame Culture - AHFE; (b) Archetype Analysis - AHFE; (c) Executive Function Analysis - AHFE; (d) Cultural Risks causes dynamic behavior - GCPS; (e) Social Risk: Social Conflicts, Accidents - ESREL. In case discussed in the Fertilizer Industry, we proposed Cognitive Academy and Habit Change Campaigns to break vicious cycle of human error.
Cultural risks also lead to inappropriate behavior. In a certain public industry, repetitive behaviors that result in low productivity, reliability and competitiveness were analyzed. These events are summarized in socio-human failure trees indicated in Figure 2.
Thus, we can build negative cultural actions, such as: following inadequate standards, generating plant stops; teams that move only for financial compensation; teams that resist change and are afraid of losing power; constantly conflicting policies and practices; personal characteristics of dispersion and depression; characteristics resulting from fear, frustration and lack of courage; when self-centeredness occurs, excessive self-confidence causes human error; in fact, production is a priority and safety - the environment as a second priority).
Vaccines for this cultural disease can be actions to break the vicious cycle; seek emotional balance; knowledge about bad habits and processes that lead to human error; neutralize political forces opposing undue behavior; awareness of the need for change; team integration; exercise the investigation of causal nexus, demanding from the organization the construction of educational campaigns and cognitive academy.
The characteristics of cultural illness as shown in Figure 2 are: low perception, improper communication, lack of true leadership, strangeness, strategy to invade personal space, simple mental map, compromise corrupted by politics. This situation demands educational-cognitive programs and cultural improvements. In the analysis for the prediction of the team's behavior and the process-activities, it is necessary to identify which cultural biases and / or possibilities of human errors are not avoided, to adjust these cognitive gaps to have resilience in the face of risk. dynamic that causes the degradation of design barriers (resistance) and the possibility of increasing the cognitive load at work. The human elements to avoid failure must be reviewed according to organizational performance threats.

2.3. Just Culture: Link between Senior Management and Control of Operational Culture

Just culture has been one of the key elements in supporting a successful safety culture in the aviation industry, in a first and most effective way, since the 1990s [17]. Nowadays, other domains such as health organizations, among other HROs, are following this same path in the search for mitigations and / or solutions to their incidents / accidents, also looking for organizational factors as possible latent causes of them.

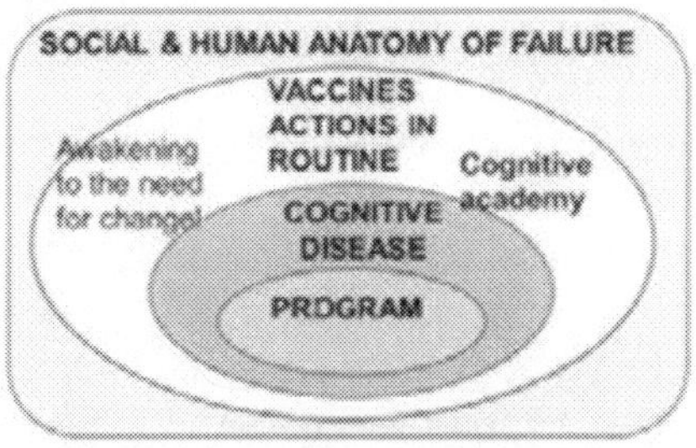

Figure 2 – Human and Social Failure Anatomy [22]

There are two fundamental aspects of the safety culture: (i) "components" or elements of its constitution and (ii) "dynamics" or aspects of interactions between its components [17]. "Components" refer to the four parts that collectively describe the safety culture: values, leadership strategies, attitudes, and performance. Dynamic, on the other hand, concerns the interactions between its constituent parts that determine the dominant or prevailing cultural state. In other words, they are the result of interactions between individuals, groups and the organization as a whole, which can neutralize problems of a latent nature that would lead to failure or that can generate resilient conditions that interrupt the trajectory of the error culminating in failure.

Patankar & Sabin [17]. also mentions a second highlight: the dynamic configuration of values, leadership strategies and attitudes that result in four "states" or dominant stages of the security culture. These states were defined as: "hiding the problem" (secretive); "The culprit" (blame); "Reporting" (reporting); and "the just" (just).

Reason [2], one of the precursors in the study of this theme, describes a Just culture as an atmosphere of trust in which people are encouraged, even rewarded, for providing essential information related to security, and where the line is clearly drawn between the acceptable and unacceptable safe behavior. Moving from a culture of guilt to a just culture, it is essential to have a robust and effective system for communicating events and reports in operation. Such a reporting system allows employees to report their own systemic errors or risks without fear of reprisals. The level of confidence of employee management in each organization is the most critical factor in determining or predicting the success of such a system.

In an evolutionary perspective among its states, the just culture differs from the culture of guilt in giving "carte blanche" to the guilty party when a certain error occurs (under a "Just Culture" condition, individuals are not guilty of "honest mistakes", but it should explain about intentional violations and growing neglect). The characteristics that classify culture as "Just" go far beyond this, since "blameless culture" could lead to unexpected events such as intentionally unfulfilled procedures or other behavior in the workplace that could affect the safety of an organization. Negatively [18].

Recent discussions bring two new aspects to the scope of research in the field of Just culture in HROs: accounting and learning. In a Just culture environment, organizations are controlled in relation to the project and in relation to incidents, not individual ones. In this way, Just culture is more than politics, it becomes a mixture of philosophy of the sense of justice with management controls [19]. It is assumed that the punitive adjustment does not solve the problems. People work within the System designed by an organization, so when an individual is at fault, the system is often at fault as well. Punishing people without changing the System only perpetuates the problem rather than solving it.

A just culture balances the need for an open and honest reporting environment with the end of a quality learning environment and culture. And this requires a change in focus from errors and outcomes to system design and management of the behavioral choices of all employees [20].

Hospital organizations show examples of a just culture that goes beyond ensuring that employees feel free to register errors. These are reliable organizations that promote full attention to their workers through practices such as: (i) a constant concern with the possibility of failure; (ii) deference to specialization, regardless of position or status; (iii) ability to adapt when the unexpected occurs; (iv) the ability to concentrate on a task while having a sense of the big picture; and (v) the ability to change and level the hierarchy to suit a specific situation.

In this way, everyone in the organization is continually learning, adjusting, and redesigning security systems and managing behavioral choices [20]. Thus, a Just Culture supports the learning of unsafe acts to improve the level of security awareness through improved recognition of security situations and helps to develop conscious articulation and the sharing of security information. Consequently, a Just Culture can be considered a facilitator, and even an indicator of (a good) Safety Culture, with the Leaders being responsible for understanding the nature and scope of errors, actively redesigning defective systems and valuing voluntary reports [21].

Discussed the new concepts that must be taken into account in the current context of the operational and safety culture, mainly in industries that require high reliability (HROs) and the dimensions and variables of what is meant by a Juster organizational and operational culture, presented a framework that will assist organizational leaders in decision making, as well as in the implementation of a more just culture based on the experiences identified and learned from the culture of their shop floor activities and processes - the operational culture.

3 Methodology

Chemical industry case studies will be used to enrich the discussion of new concepts and the constituent steps of the proposed framework.

The construction of this framework is the result of the appropriation of previous experiences in the development and validation of concepts, methods, tools and applications of real cases in the area of Human Factors, Risk, Reliability and Crisis, carried out by the research group GRODIN (Group of Operational Risks and Dynamic due to the variability of behavior and external environments), throughout its 10 years working in several industrial sectors (food, fertilizers, refining, LPG gas and petrochemicals) and services (mega events, fire brigade and civil defense in the State of Bahia). The refinement of the framework also counted on the participation of several specialists with great experience in production and security management. In the same way, managers, and supervisors of production plants in the mentioned industrial sectors were also consulted, through opinions, in contribution to the improvement of the proposed framework.

The framework proposal (figure 3), which aims to migrate from a guilt culture behavioral context to a more just culture, is presented below and was developed considering three layers of investigation and intervention: diagnosis; post-diagnosis; and continuous learning.

These layers are integrated and obey the principle of continuous improvement (PDCA cycle). It is also worth mentioning the minimum estimate of 3 years to start achieving the desired results, since habits and behaviors in constitution and change of a state of culture are a function of time variable.

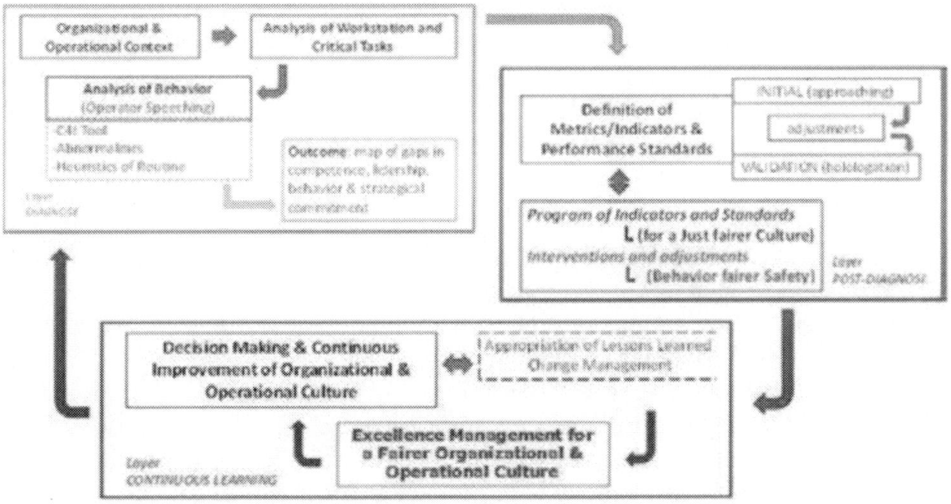

Figure 3 – Framework proposal: migration from blame to just culture.

4 Case: Chemical Industry, Oil and Gas

The chemical industry, as well as the petrochemical, a nuclear and the aviation sector, are called high-trust industries (HROs). In this industrial segment, the high complexity of its processes raises the challenges with which the management of its proposals manages and improves the elements of the organizational project (structure, design, procedures, attitudes and behavior of the leadership and the shop floor) that promotes better performance. , not only in terms of safety but also in operational efficiency (and in the quality of its manufactured products) and in the results of environmental and sustainable performance. At the Workstation there are several indicators regarding the distance or the approach of the Just culture. The high turnover of the workforce in the production teams can indicate the lack of commitment to the company and to safety, being, therefore, a variable that indicates the level of risk in these complex industrial processes.
Risk perception is a characteristic that integrates the safety culture, the operator's behavior, and the cognitive quality of the team. The low perception of risk indicates unsafe behavior resulting from the operational routine. The low commitment to the quality results of the resulting products on the part of the work teams, impacts their efficiency (the operational costs are increased in value and in rework operations) and in the motivation of the leaders and supervisors in investigating the causes and seeking solutions. The organizations of the chemical sector, have their industrial processes very varied, having tubular reactors of fixed bed, distillation at low pressure, evaporation, absorbers, crystallization, dryers, centrifugation, among others. Some of its biggest risks are the storage of both flammable and toxic raw materials and finished products; atmospheric emissions; and the release of contaminated liquid effluents.
Experiences carried out in different facilities to achieve Just culture are discussed to confirm the steps of this methodology as indicated in Figure 5 in A: (1) Chemical Industry Indicators (formaldehyde) for Just Culture; (2) Diagnosis of Human Factors in the LPG Industry Task; (3)

Intervention with an educational program to avoid contamination of the effluent with ammoniacal nitrogen in the Fertilizer Industry; (4) Improvements in the Polycarbonate Industry to reduce contamination of the effluent with methylene chloride; (5) Analysis of Communication Quality in the onshore oil production industry.

Relating the activities with the steps of the Just Culture Method, we can affirm that the analysis of the operational context with the identification of deviations and types of human error, in order to arrive at the diagnosis, by mapping abnormal events. Interventions after measuring human factors and indicators, which signal the direction for a just culture (post-diagnosis). And the standardization of changes, to achieve a Just culture and a new operational context (safer behavior). In the current description, we will develop only the case of the chemical industry and the flammable gas industry.

4.1. Measurement of direct and indirect Human Factors in the chemical industry

In a case of the chemical industry, indicators of the task (for example specified quantity produced), culture (with number of deviations in the shift or related deviations in maintenance, such as disruption in the sealing of rotating equipment influencing the contamination of organic effluent) and behavior (level of stress caused by% doubling or overtime, and presented by bioindicators of the average type of blood pressure, presence of gastritis and ulcers, and in the weight-for-height relationship, these indicators, in the long run, may indicate the direction of just culture.

This industry produces 37% formaldehyde to produce synthetic, phenolic and urea resins used in the sectors of wood, plastics, paints, varnishes, paper, cellulose, textiles and foundries. The task indicators for safety, environment and production show better results in the 2017 to 2020 series, requiring a reduction only in the hour man at risk. These are examples of indicators for achieving a Just Culture.

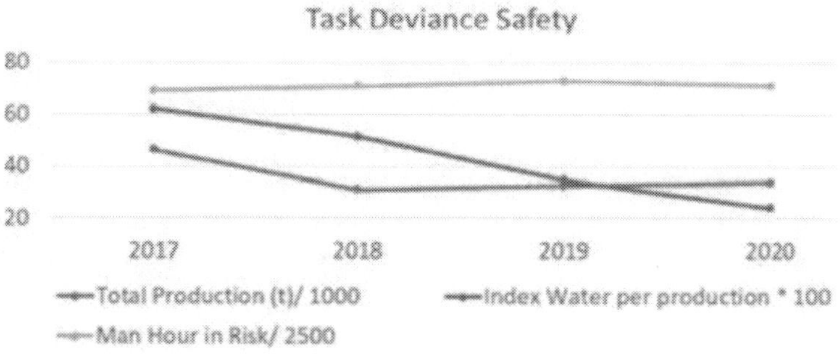

Figure 4 – Efficiency Task Indicator – Safety, Environment and Production

4.2. Measurement of direct and indirect Human Factors - LPG

The scheme in figure 5B, presented below, illustrates the path of the analysis for a future situation of organizational efficiency in carrying out the task; emotional balance related to stress and decision stability; aggregation and disaggregation of the team reflected by questionnaire and scenarios - direct reflection on cooperation; and level of commitment.

The Tool C4t identify behavior from features in the work with self-response about the routine. We presented the classes of Human Factors from 1-9 in 2.1 to explain systemic failure. But now we are talking about measurement of human elements in case of LPG site part of SARS Tool [6]. Some indicators were tested in Oil & Gas and Petrochemical Industry. We understand that emotional state (Figure 5C) is based on cultural aspects, cognitive quality, and possibility of psychosomatic diseases, giving in Gas Industry a final value of 7 from 10 and standard 8. In case of commitment, we consider worker information as time at company, if married and how many children, and other personal information. We tested giving as result 3 groups were 30% in high commitment level (HE5). The level of cooperation is based on aggregation (AG) and disaggregation (DAG) parallel movements, in the Oil and Gas indicating, AG = 6,8 from standard

8, DAG = 5 from standard mean of 5,4 (HE6). The competence analyzes is based on the human errors during long period of shift, 2-6 months, and task results. In this period, we can elaborate the cognitive and behavior gaps and create an exam that will be applied in operational group and indicate the training and behavior necessities to avoid the accidents (HE7). Finally, Ávila and Drigo [13] elaborated Organizational Communication Model to measure the quality (omissions and commissions) of ascendent, horizontal and descendent communication (HE8).

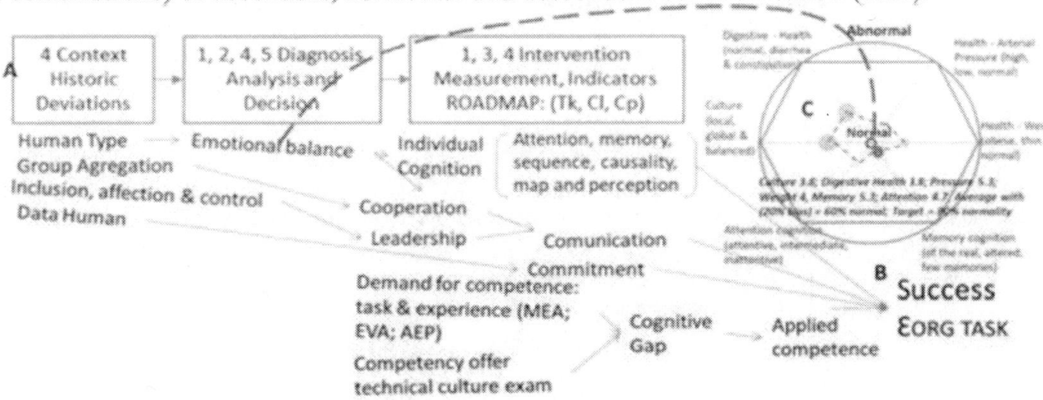

Figure 5 – Diagnosis of success in the task and organizational efficiency [6]

5 Final Considerations

The proposed framework is the result of the background and expertise built by GroDin's research activities in its 10 years of existence. This framework incorporates new concepts such as systemic failure; cultural force field; socioeconomic and affective utility of the task versus feedback; causal link of low-risk perception; difference between enabling elements of hazard energy versus human factors in the operational situation; management models to avoid accidents and to avoid energy losses and respective techniques that include the discussion about culture operational and human factors.

In its integrated structure - chains of its activities or constituent parts - it allows a natural transition between the layers presented, enabling continuous and evolutionary feedback flows regarding improvement and learning. The starting point (diagnostic layer) is the identification of gaps in skills, leadership, behavior and strategic commitment, through the investigation and analysis of critical tasks, human elements and factors and behavior via the operator's speech. The definition and validation of metrics, standards and adjusted indicators are core activities of the framework that impact, with greater effectiveness, the entire intervention effort towards a migration from a cultural state of inefficient performance, from uncompromised and insecure workforce behavior; for an evolutionary state of integrated performance of the organizational functions, including productivity-quality of the products and processes, level of security of the workstation the plant as a whole, and of the best efficiency in the use of the natural resources, as well as in the control of mitigation of its effluents. A more just culture achieved results in an atmosphere of trust between individuals, teams and leadership levels, which enables adjustments to the organization's learning process: appropriation of lessons learned, processes of reporters (accidents / incidents) and feedback and change management (from operational to organizational).

The cases presented are examples of empirical evidence already tested in the chemical, petrochemical, and LPG gas industries and which validate the framework proposed in this paper.

References

[1] J.A. Klein and B.K. Vaughen. "Implement an Operational Discipline Program to Improve Process Plant Safety," Chemical Engineering Progress, pp. 49-52, June 2011.

[2] Reason, J. (1997). Managing the risk of organizational accidents. Aldershot, UK: Ashgate.

[3] Ávila Filho, S., Sousa, C. R., & Carvalho, A. C. (2015). Assessment of complexity in the task to define safeguards against dynamic risks. Procedia Manufacturing, 3, 1772-1779.

[4] Ávila Filho, S., Cerqueira, I. C., Ferreira, J. F. M. G., & Drigo, E. (2018, April). Task & Communication in Just Culture, OIL Offshore Exercise. In 2018 Spring Meeting and 14th Global Congress on Process Safety. AIChE.

[5] Ávila, S., Mrugalska, B., Ahumada, C., & Ávila, J. (2010). Relationship between human-managerial and social-organizational factors for industry safeguards project: dynamic Bayesian Networks. In 22nd Annual International Symposium Mary Kay O'Connor Process Safety Center, College Station, Texas.

[6] Ávila Filho, S. Etiologia das Anormalidades Operacionais na Indústria: Um modelo para Aprendizagem. 2010. Tese de Doutorado. Universidade Federal do Rio de Janeiro. Escola de Química, Rio de Janeiro.

[7] PERROW, C. Normal Accidents: Living with High-Risk Technologies. New Jersey: Princeton University Press. Basic Books, NY, 1984.

[8] Righi, A. W., & Saurin, T. A. (2015). Complex socio-technical systems: Characterization and management guidelines. Applied ergonomics, 50, 19-30.

[9] Saurin, T. A., Righi, A., & Henriqson, É. (2013). Characteristics of complex socio-technical systems and guidelines for their management: the role of resilience. In 5th Resilience Engineering Association Symposium, Soesterberg, Holanda.

[10] S.A. Filho. "Reliability Analysis for Socio-Technical System, case propene pumping." Engineering Failure Analysis, 56, p. 23, 2015.

[11] S.A. Filho, D. Santos. "Integrated reliability: failure assessment to process, operation, equipment and human integration (POEHint), a case of propene pump operation." Sixth International Conference on Engineering Failure Analysis, Lisbon, 2014.

[12] S.A. Filho, C. Dias. "Reliability research to design barriers of sociotechnical failure." European Safety and Reliability Conference, pp. 345-345, 2017.

[13] Ávila, S. Methodology to minimize effluents at source from the investigation of operational abnormalities. Dissertation (Professional master's) – TECLIM, UFBA, Salvador, 2004.

[14] S. Avila, J.S. Ávila, L.M.Pereira and E.R.F.Lima. "Dynamic View of Human Elements Design and Projection of Human Factors in Critical Task Operation Using Bayesian Networks". 2020 International Conference on Applied Human Factors and Ergonomics. July 16-20th, 2020.

[15] A.C.F. Carvalho. "Diagnóstico de lacunas nas competências da equipe para enfrentamento de situações fora da rotina". Master's dissertation – Federal University of Bahia, Polytechnic School of Engineering. Salvador, 2016, 103 p.

[16] S.Á. Filho, R.A. Kalid, A.C.F. Carvalho and R.C.O. Sousa. "Influência da Cultura Brasileira no Comportamento do Trabalhador e na Disciplina Operacional". In: ABRISCO (Brazilian Association for Risk Analysis, Process Safety and Reliability) PSAM, 2015.

[17] Patankar, M. S., & Sabin, E. J. (2010). The safety culture perspective. In: Human factors in aviation (pp. 95-122). Academic Press.

[18] Eurocontrol (2006). Establishment of Just Culture Principles in: ATM Safety Data Reporting and Assessment. ESARR Advisory Material/guidance Document.

[19] Paradiso, L & Sweeney, N. (2019). Just culture: it's more than policy. Nursing management, 50(6), 38.

[20] Boysen, P. G. (2013). Just culture: a foundation for balanced accountability and patient safety. Ochsner Journal, 13(3), 400-406.

[21] SKYBRARY – Just Culture. https://skybrary.aero/index.php/Just_Culture (accessed March. 10, 2021).

[22] Ávila, S.; Silva C.C.; Souza R.C.O. Worker behavior changing by dynamic culture. 9 GCPS. San Antonio. 2013

619665 The DO'S and DON'TS of Distillation Control — The Practitioner's Perspective

Abstract:

Product separation using Distillation is still relevant today. Distillation control does not appear to be an issue anymore, judging by the number of recent publications in the field. However, problems with distillation proper operation and control are still latent in the chemical processing industry, a statement that cannot, unfortunately, be supported by any statistics or study, as no production facility would be willing to disclose any information to the public on deficiencies or problems with their operation.

With several distillation columns being operated in manual or with a single composition control, and most of the nowadays efforts in the area of industrial automation being focused on the tools of the fourth Industrial Revolution (4IR), process control appears to have been moved to the back seat; whilst the process control professionals on the ground are not able to get much support as the experts retire or abandon the field, a topic that is outside the scope of this writing.

Given the benefit of its contribution to the practitioner, this paper offers hands-on recommendations on what should and what should not be done, when it comes to distillation control. Recommendations that are based on the author's years of applied experience on successful delivery of distillation control strategies.

Some of the recommendations included in the paper are: Is there a one-fits-all configuration? When is material balance control the most indicated? What temperature should be used as a composition surrogate? Does Pressure Compensated Temperature always work? Is Model Predictive Control the best approach to Distillation Control?

619666 Improved Distillation Composition Control

Abstract:

The traditional means of controlling a distillation column product composition is via a PID-to-PID cascade, analyzer to temperature. Another approach is to use Model Predictive Control (MPC) to control the product analyzer value with a column temperature controller as the primary manipulated variable, i.e., with the temperature controller in cascade to the MPC.

Natural questions arise: "Are there simpler control options when an MPC is not justified, e.g., when constraints are not a significant issue" and "Can we do better than an analyzer to temperature cascade?"

As with any cascade loop, there is the requirement that the inner, secondary loop (the temperature controller) be sufficiently faster than the primary loop (the analyzer controller). Unfortunately, the dynamic response between the manipulated variable (or disturbance) and temperature and between the temperature and analyzer are not sufficiently different, making it difficult to achieve this requirement in practice (without significantly detuning the analyzer controller). As a result, when a composition disturbance occurs, the temperature controller responds first to correct for the temperature deviation and when the analyzer responds later (due to slower dynamics and analyzer cycle time) the analyzer controller responds by changing the temperature setpoint, which if often unnecessary and prolongs the return to steady state.

An alternative approach is to develop a model between the controller temperature PV (not SP) and the analyzer PV. In using this model for the controller the temperature SP is only changed when model offset occurs or when the analyzer SP is changed. The model and controller, which is akin to internal model control, can be implemented fairly simply in a DCS or PLC.

The details for developing and implementing this scheme are described along with guidance for selecting the control temperature(s).

619672 Effect of Blanking on the Performance of Sieve Trays

Abstract:

Blanking is a common practice used to modify tray decks to obtain satisfactory operation at vapor rates lower than those possible using the normal or full perforated area of the tray. Blanked trays may be found in new towers designed for anticipated future capacity expansions or in revamped towers, especially when their new or current operational conditions are very different than those used for the original design or when the columns are repurposed for different services.

Throughout the years, FRI's legendary expert Tak Yanagi conducted a series of studies aimed to determine the effect of using blanking strips on the hydraulic performance and efficiency of sieve trays. During such studies, changes in design and operational parameters have been analyzed: orientation and location of the blanking strips, hole size, hole area, operational pressure and test system. This presentation will provide an interpretation of Mr. Yanagi's findings about the effects of using blanking strips in sieve trays.

619667 Droplet Formation and Reduction in Mass Transfer Equipment

Abstract:

This contribution presents scope and major results of the joint research project "TERESA – Droplet generation and reduction in mass transfer equipment" financed by the German government including 4 universities and 8 industrial partners and developed over 4 years.

Mass transfer and separation processes are the main unit operations in the chemical, petrochemical, pharmaceutical and food industries. Mass transfer based on evaporation and condensation is very energy intensive. It is estimated that approximately 3% of the world energy consumption is attributed to thermal separation processes. These rely on the continuous establishment of vapor/liquid equilibrium followed by a perfect phase separation. This research project centers on the separation of liquid and gaseous phases after a mass transfer step: entrained droplets must be separated from the gas phase and gas bubbles from the liquid. Entrainment and back mixing reduces separation efficiency, causes higher energy demand and can lead to corrosion and process stability problems. To avoid such troubles, a better understanding of the formation, entrainment and separation of droplets is essential.

The project is focused on four major trouble areas: the feed line of a flashing feed, the feed section inside the column, the two-phase flow of a forced circulation flash reboiler and the head vapor of a distillation column. Flow pattern maps are common to characterize the two-phase flow in the feed pipe. These maps are used and were optimized to larger pipe diameters, and typical but non-ideal industrial conditions.

To detect droplets (diameter > 2 micrometer) and to measure gas/liquid ratios existing devices (capacitance-wire-mesh-sensor, optical inline probe) were modified and adapted to the conditions and calibrated. These highly sensitive devices were used to measure droplet size, droplet concentration and gas/liquid ratios in the feed line, as well as inside an air/water column (e.g. between two trays) and the gas outlet. To adapt air/water tests to organics, investigations were made with organic mixtures in a reboiler line and after expansion in a feed line. Integral entrainment data were compared to local droplet measurements and showed excellent agreement in the quantification of critical operating parameters.

Based on these experiments improved droplet separation devices were implemented and tested. In addition, CFD calculations were used to get a better understanding of flow patterns and to improve design criteria for the development of droplet separation devices. Economic and ecologic investigations confirmed the advantages of improved droplet separation devices minimizing entrainment in mass transfer equipment.

619158 Troubleshooting of a Caustic Scrubber in a Chemical Plant

Abstract:

A caustic scrubber in a chemical plant has occasionally not been able to meet the target HCl concentration in the vent gas stream. To avoid a reduction in production rates or a complete shutdown of the plant, a rate-based simulation of column was developed and used to investigate potential column performance limitations and identify areas of improvements. The column capacity was evaluated, gamma scans were analyzed, and possible aerosol formation was investigated.

ELIMINATE BOTTLENECKS - DESIGN EFFECTIVE CONDENSATE DRAINAGE FROM REBOILERS

VIRTUAL
SPRING21
+17th GCPS
A Joint AIChE and CCPS Meeting
APRIL 18-23, 2021
Securing the Future through Chemical Engineering & Process Safety Innovations

JAMES R. RISKO, CEM, PEM, MBA

ELIMINATE BOTTLENECKS - DESIGN EFFECTIVE CONDENSATE DRAINAGE FROM REBOILERS

JAMES R. RISKO President TLV CORPORATION Charlotte, NC

Reboilers using an inlet steam control valve (ISC) arrangement can provide significant economic and control benefits to certain distillation tower installations. By exposing the full surface area of a reboiler to steam for effective heat transfer, the steam pressure can be lowered - which can reduce the fouling that can occur with higher pressure steam and the corrosion that can follow from flooding a tube bundle. Control can improve because steam can adjust to process demand changes much more rapidly than by the alternative of varying the condensate liquid level and exposed area of the tube bundle - such as occurs with an outlet condensate control valve (OCC) arrangement [1]. Furthermore, unlike OCC, it is generally not possible to lose the condensate seal/reboiler duty when the proper drainage design is implemented. Other potential benefits include mitigation of the stratification that can plague channel head gasketing in horizontal designs and enable the use of "long" low pressure steam to improve the steam balance or reduce generation requirements for medium pressure steam.

The Issues

In order to achieve those aforementioned benefits, reboilers with ISC have to overcome certain potential operational pitfalls, which can create bottlenecks in the system drainage and hinder process operation. These include:

- insufficient pressure to discharge condensate into the return header on start-up, under low demand, or when the tube bundle is relatively clean.
- design uncertainty relative to the drainage system utilized – steam trap (and if so, which type?), pump/trap, level pot, electric pump
- insufficient understanding of balance requirements for level pots or pump/trap reservoirs that can lead to reboiler evacuation issues, such as stall, backup, hammer, and corrosion.

Effective reboiler system reliability requires timely, almost simultaneous condensate drainage from the steam space. Unfortunately, there are multiple instances of ISC and OSC reboilers experiencing condensate back-up and related issues such as channel head gasket leaks, tube bundle damage, and sewering of condensate. These issues may be commonly experienced in both horizontal and vertical reboilers, with horizontal systems particularly challenged by channel head gasket leaks. (Figure 1).

Many of these problems can be avoided/mitigated with a proper condensate drainage design, but it often occurs that the reboiler system experiencing difficulty had improper original design or installation and requires correction for improvement. As an example, it is typical that a level pot arrangement may be improperly balanced, or a stall condition may not be recognized and therefore, not mitigated. Additionally, some reboiler operations result in the need to discharge colder, non-flashing condensate into headers containing flash steam, a process that can produce violent water hammer in the return line.

It is possible to explore historical negative experiences, analyze why certain problems occur, and provide some specific key design considerations necessary to obtain optimized reboiler performance.

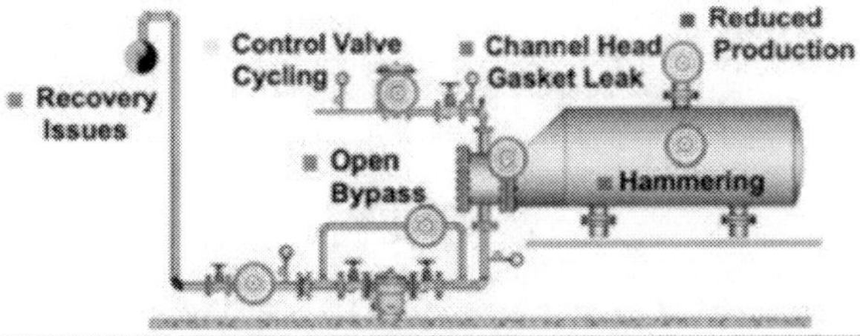

△ **Figure 1.** Improperly drained reboilers can bottleneck the system and experience multiple reliability issues.

Reboiler Basics

ISC Reboilers commonly drain condensate through a steam trap on the outlet side (Figure 2). This arrangement can provide mixed results depending on the range of differential pressure across the steam trap. Consider the pressure before the ISC as "P1,", the pressure exiting the control valve and entering the reboiler as "P2,", its exit steam pressure at the inlet of the steam trap (plus head) as "P3,", and the back pressure against which the trap must discharge as "P4." The differential pressure across the trap is expressed as "P3 - P4." If the differential pressure is positive, and "P3 - P4 > 0," then just a steam trap is needed to drain condensate. But, if the differential pressure is negative, there can be instances of undesirable condensate back-up into the reboiler, which will be subcooled by the process liquid. When steam contacts the subcooled liquid, shock damage from collapsed steam is a common result. A question may follow, "How does the differential pressure become negative if the P1 value is much higher than P4, such as 150 psig (P1) to 20 psig (P4)?"

△ **Figure 2.** Reboilers with inlet steam control typically drain condensate overhead through a steam trap.

It is worthwhile to review the reboiler heating process, and the control systems used to equalize the supply heat to the demand heat. Supply heat has to equalize with demand, otherwise the liquid will be over-reboiled or under-reboiled. There are 2 main control schemes, ISC or OCC. (Figure 3).

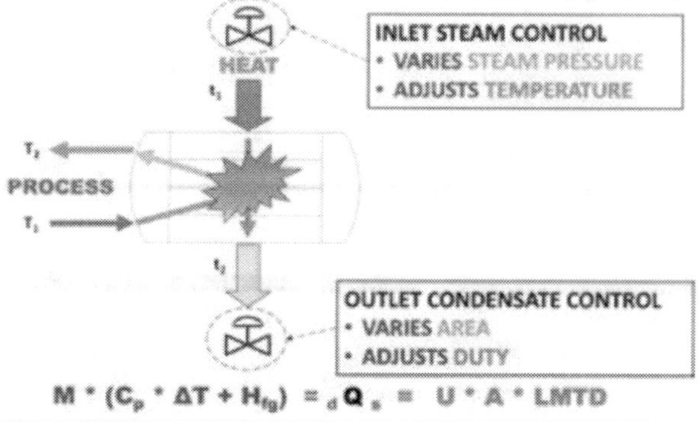

△ **Figure 3.** Effective heat transfer requires that control schemes provide supply heat that can equalize to heat demand.

The OCC method varies the reboiler's unflooded area to adjust the duty. The flooded area does not transfer heat of vaporization, but only provides heat by subcooling of the condensate. This heat exchange by sensible heat has a much lower heat value and transfer coefficient than by steam heat exchange. So, an OCC reboiler's heat duty is manipulated by varying the unflooded steam-condensing area. OCC systems in horizontal or vertical systems can experience trouble with instability/operational issues, tube corrosion, or steam blow-through, but may experience less difficulty from stratification and hydraulic/thermal shocks in vertical shell-side steam reboilers than in horizontal reboilers. The higher steam pressures of OCC systems may limit process fluid backflow into the steam line from a tube leak [1], but elevated temperatures on the tube set may increase fouling of certain temperature sensitive liquids. It is usually easy to determine if a system has experienced difficulty because there are commonly visible signs. (Figure 4).

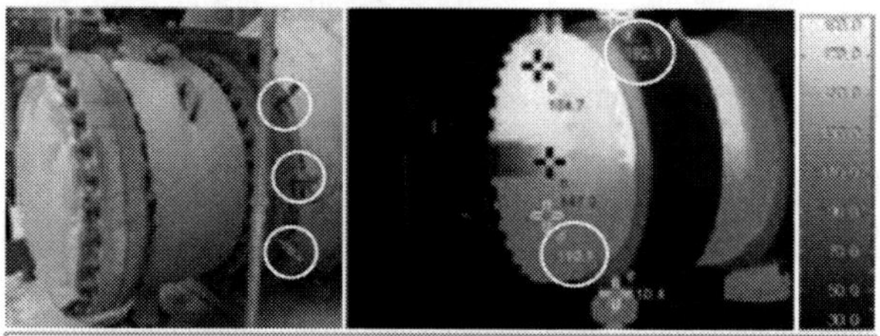

△ **Figure 4.** Leak collars can indicate temperature stratification and condensate back-up issues.

ISC systems adjust supply heat to demand by varying the steam pressure, which adjusts steam temperature and the mean temperature difference between heat and product flows. This allows the lowest pressure steam – often vent steam in a rebalanced system – to be used. Lower pressures of ISC schemes may allow process contamination into the heating system from a tube leak [1]. The process fluid backflow into the steam supply line can be mitigated by a check valve installed in the steam line after the ISC and before reboiler entry, but process contamination of condensate still will occur. Whether using medium or low-pressure steam, the ISC arrangement can reduce steam pressure into vacuum in order to equalize the supply heat to demand, particularly when the tube bundle is unfouled and the demand is not at peak. When the load is near peak or when the tube bundle is sufficiently fouled, higher pressure steam can be present, and a positive pressure differential can exist across the steam trap. This enables steam to be prevalent throughout the tube bundle (Figure 5).

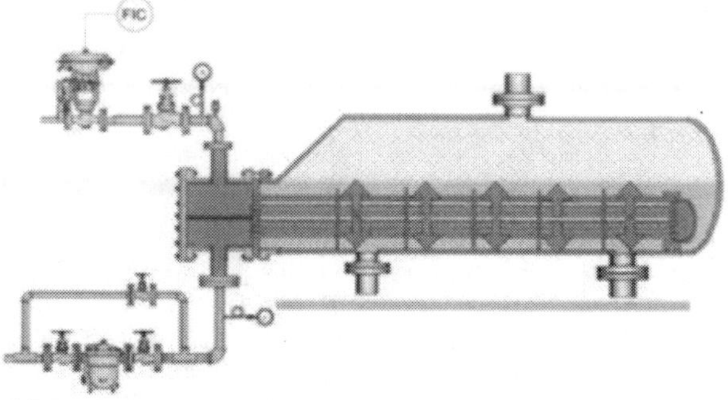

△ **Figure 5.** Reboilers that heat without condensate back-up experience fewer maintenance and control problems.

But, when demand load decreases or when the tube surface is relatively clean (and over-surfaced), there is so much heating area capability that the ISC throttles to reduce the steam pressure and corresponding temperature. At a certain threshold of reduced load (in either horizontal or vertical reboilers), the steam pressure will equalize with the back pressure, creating a "stall point." The differential pressure across the steam trap is 0 psi, so no flow can occur for that heat demand or lesser. The condensate flow is "stalled" inside the equipment. When a stall occurs, condensate backs up into the reboiler and the condensate becomes subcooled, creating corrosion and the possibility of hydraulic shock from a phenomenon known as "collapse hammer." This situation can also cause significant channel head gasket issues in horizontal equipment (Figure 6).

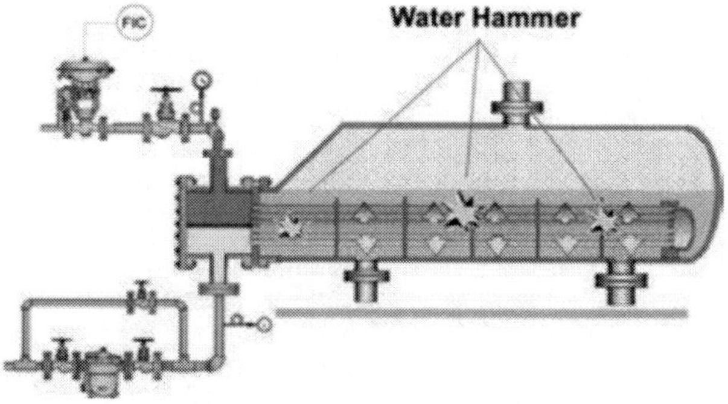

△ **Figure 6.** Reboilers that experience condensate back-up can suffer water hammer shock and resulting issues.

Extended Stall Chart Load Analysis

As heat load demand reduces, the steam pressure throttles to equalize supply heat to demand. A stall point occurs when the exit pressure, P3, falls and equals the condensate side back pressure P4. It can be helpful to use the Extended Stall Chart

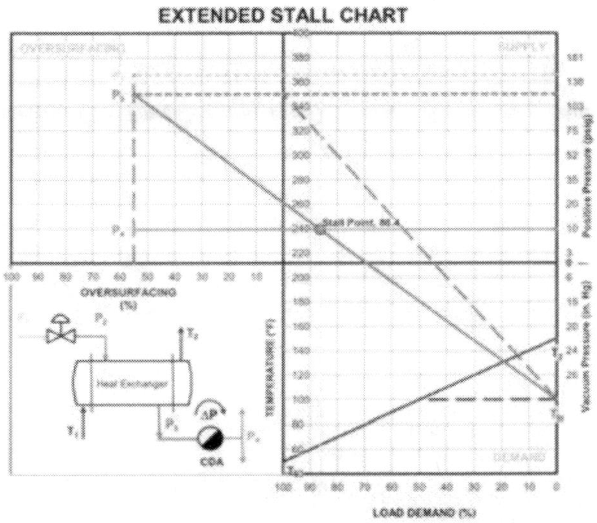

△ **Figure 7.** When an over-surfaced tube is clean, against low back pressure (10 psig) stall can occur at 86.4% of full load.

[2] to estimate various load versus back pressure scenarios. Three hypothetical instances plotting the percentage of demand heat load at which the supply steam pressure intersects with the condensate back pressure are considered:

1. 150 psig P1, 10 psig P4, **55%** over-surfacing (Figure 7)
2. 150 psig P1, **20** psig P4, **0%** over-surfacing (Figure 8)
3. 150 psig P1, **20** psig P4, **55%** over-surfacing (Figure 9).

Figure 7 shows an estimated stall point at 86.4% of full load when discharging against low back pressure of 10 psig. As long as the load is ALWAYS above 86.4% of full load, it is possible to design drainage using just a steam trap with a check valve on its outlet (to mitigate back slam damage to the trap – Figure 2). This is the lowest cost installation and can be effective where there is little load variation.

In Figure 8, 2 dynamics have changed, 1). The equipment has fouled – having the same effect as moving the area to just meet 100% of demand (no over-surfacing). 2). The P4 system back pressure has increased to 20 psig and the stall point has actually reduced to 63.5%. When the tube surface is fouled, it can become more difficult to experience a stall condition, and the process can perform well for multiple months against the higher back pressure. That situation can cause a production unit to not apply sufficient focus to keeping the back pressure low, but that can become a contributing factor to future stall conditions and equipment damage issues.

Figure 9 explains by showing the same system AFTER the tube bundle is either cleaned or replaced. With the added "UA" capability of a clean surface, the P3 exit pressure equalizes with the elevated back pressure (P4) at a stall point of 98.4% of full load. The equipment can experience stall anytime the load is not at maximum. The original stall point was 86.4% of load, but the increase of just 10 psi back pressure (to 20 psig) caused the stall point to move to 98.4% of full load. This analysis reflects the crucial importance of keep condensate system back pressure low.

There are typically four reasons why back pressure increases from the original design:

- Open bypass on some equipment connected to the same condensate header
- Steam trap population is not proactively managed and is experiencing a lot of steam blow-through into the condensate return header [3, 4, 5]
- OCCs on some heating equipment are suffering from steam blow-through [1], particularly on those systems which do not have level pots installed
- System dynamics have changed, e.g., a new condensate tie-in with significant added flow has connected to the header.

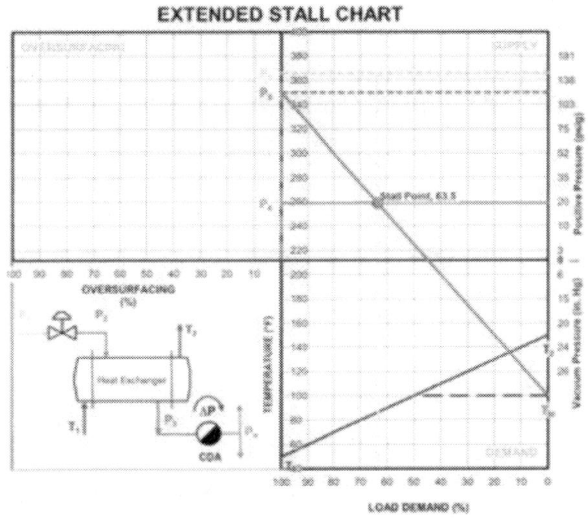

△ **Figure 8.** When a tube surface is fouled, against elevated back pressure, stall can occur at a lower 63.5% of full load.

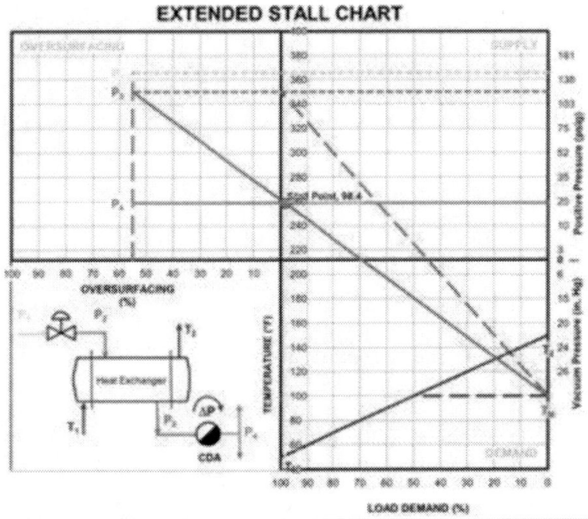

△ **Figure 9.** When an over-surfaced tube is clean, elevated back pressure increases the stall point to 98.4% of full load.

Design for Handling Stall

A stall condition can be misunderstood as something to be avoided, but that may not reflect the opportunities that such a condition affords. Conditions which encounter stall enable the lowest possible steam pressure to be used, and often allow for low pressure vent steam to perform the heating. With the high steam flow demand of certain reboiler operations, this can represent a significant steam balance improvement, and a lower steam pressure can reduce fouling of temperature sensitive product streams. Horizontal reboilers may experience channel head gasket damage from the flooding of equipment done by OCC schemes, and because ISC methods heat by vapor, they can provide better reliability as long as the condensate is quickly drained.

While ISC enables rapid adjustment to demand changes, the lowered pressure and possible negative pressure differential across the outlet drainage device may not permit condensate to be discharged into the header, so a method to pump the condensate into the higher back pressure and capture the benefits of ISC operation can be needed.

One method is to install a separate pump after the steam trap. While the differential pressure across the trap may be too low to discharge condensate 35 feet overhead into a return header of 20 psig pressure or greater, there is most likely sufficient differential to discharge 2 – 3 feet overhead into a condensate flash receiver and then pump the condensate overhead. Condensate in such close proximity to the condensing source and flash vessel can be close to 212°F, so a non-electric, secondary pressure drainer pump (SPD1) pump is preferred to avoid cavitation (Figure 10). Depending on the amount of flash steam and production unit location, a method to handle flash – such as a vent condenser – may be needed.

Δ **Figure 10.** To discharge against higher back pressure, one option uses a flash receiver and pump after the steam trap.

Rather than install a separate steam trap and condensate pump to discharge against a high back pressure, another design is to use a combination pump and trap in the same body, a "pump/trap" (SPD2 - Figure 11). For draining condensate from reboilers, the loads can be high so it is common to have a dual orifice trap arrangement which can be seen in the lower left of the cutaway image. The float assembly is a mechanical actuator – when high level is reached it simultaneously opens a motive valve to input high pressure "motive" steam which discharges the condensate against high back pressure and closes the exhaust valve so that the body can be pressurized. At low level, it closes off the motive valve and opens the exhaust to allow the internal vapor to balance with the condensing source – where it still has heat to contribute to the heating process.

A pump/trap is connected to the reboiler condensing source in two locations.

1. The reboiler condensate outlet supplies liquid condensate directly into the SPD2
2. The SPD2 exhaust is balanced to an appropriate outlet side location on the reboiler which will have the SAME pressure AND allow vapor to reach the tube bundle.

Δ **Figure 11.** A pump/trap assembly uses the float as actuator on the lower trap unit and upper secondary pressure/exhaust valves.

This second point is critical so that the SPD2 becomes the lowest component of the reboiler – at exactly the same pressure, to enable the gravity flow of condensate into the pump/trap and at the same time allow the vapor to reach the tube bundle to be properly condensed. If the SPD2 is improperly balanced, it can have the same effect as stuffing a rag into the exhaust of a car, causing the car (or the pump/trap) to stall. There must be a free exchange of exhausted vapor into the condensing source at its outlet pressure. This is one of the reasons why it is required to connect to a very specific balance point in order to obtain proper operation [6]. Figure 12 shows the design layout of a pump/trap assembly draining a kettle reboiler, with gravity flow from the reboiler outlet into the reservoir, and a balance to the outlet side of the reboiler

Δ **Figure 12.** To discharge against higher back pressure, a combination pump/trap assembly eliminates local external flash.

channel head. The pipe reservoir immediately above the SPD2 is different from a flash vessel in that it does not produce any flash due to its direct connection to the condensing source.

△ **Figure 13.** Low condensate outlet, tube-side steam reboiler must be balanced directly to the outlet side of the channel head.

The simplest way to install a pump/trap package system is to have the reboiler condensate outlet flange 70 - 80 inches above grade. This enables gravity flow into the SPD2 reservoir and can reduce the number of pump/trap units required. However, in some instances, particularly existing systems, the condensate outlet piping may be close to grade. It may still be possible to use an SPD2 assembly with a specialized, low fill design (Figure 13). If the pump/trap assembly cannot be utilized in ultra-low installations, then a separate steam trap and SPD1 system is used-still enabling heating with low-pressure steam, mitigating stall, and efficiently returning condensate.

Balancing a Tube-side Steam Reboiler

Case histories exist that demonstrate multiple imbalance issues of level pot balancing with both OCC or ISC schemes that use outlet control valves for draining the reboiler. Similar balance errors have occurred with incorrect SPD2 installations. Three possible level pot/reservoir balance locations follow:

1. Before ISC
2. After ISC, but before reboiler
3. Reboiler outlet-side steam space.

Consider hypothetical pressures before (150 psig) and after (120 psig) the ISC, and at the reboiler outlet (115 psig). In Figure 14, when the yellow balance line is connected to the ISC inlet side, 150 psig steam pressure carries into the level pot/reservoir. After incurring pressure drops through the ISC and the reboiler, even the reboiler's highest outlet steam pressure of 115 psig (before throttling for reduced demand) cannot enable condensate flow into a vessel holding 150 psig pressure. It would require substantial, non-existent vertical drop to create sufficient hydraulic head filling pressure. The result is significant condensate back-up into the reboiler.

Figure 15 demonstrates the various pressures when the yellow balance line connects the level pot/reservoir to a location after the ISC but before the reboiler. In the example, the reservoir pressure of 120 psig is higher than the reboiler outlet pressure of 115 psig. Although this situation is not as damaging as the scenario in Figure 14, it does present a negative pressure differential that results in back-up into the reboiler [6]. Only a vertical drop of over 12 feet could enable flow into the reservoir and typically this does not exist.

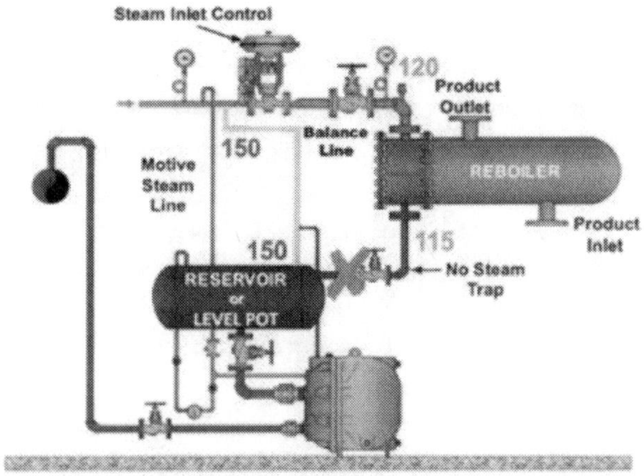

△ **Figure 14.** Reboiler outlet pressure is too low (115) to drain condensate into the level port or reservoir (150).

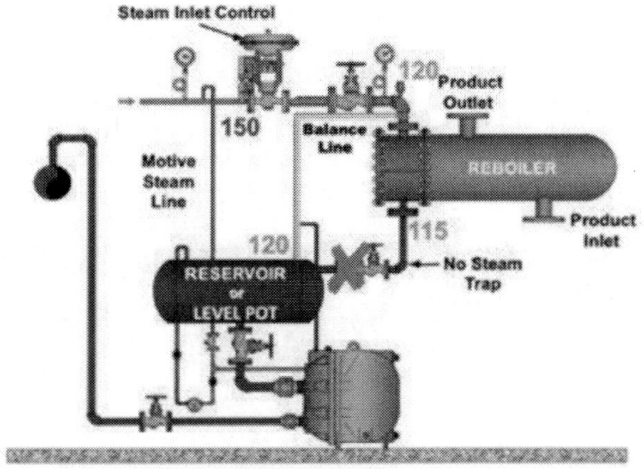

△ **Figure 15.** Reboiler outlet pressure is too low (115) to drain condensate into the level pot or reservoir (120).

Figure 16 shows the proper balance location to the vapor space which is at the same pressure as the condensate outlet connection. For tube-side steam reboilers-typical to horizontal designs-when the elevation difference between the reboiler and the condensate drum level is small-the connection MUST be to the channel head in order to obtain both a proper balance pressure equalization AND a vapor space for the exhaust steam to condense in the tube bundle space. Due to an undersized tapping on the channel head, some engineers try to balance to the reboiler outlet piping, but that normally does not work. The reason is that the condensate outlet piping is not large enough or with low velocity to enable exhaust steam to counterflow up into the tube bundle. As a result, the channel head location on the outlet side of the diverter plate is the required location, and that tapping must be of sufficient size so as to not restrict balancing. The vapor space is an excellent location to include an air vent/check valve combination to enable venting-while restricting the inflow of air (vacuum condition).

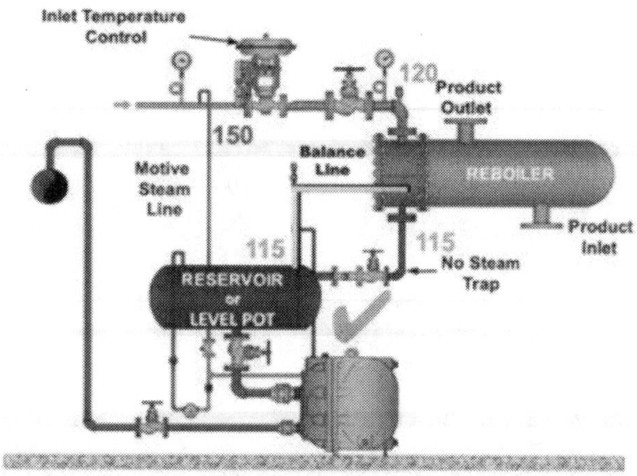

Δ **Figure 16.** Reboiler outlet pressure is equal to the level pot or reservoir (115), thereby enabling condensate to gravity drain.

Key Balance and Motive Sizes

Although each manufacturer may have different requirements for their SPD2 systems, Figure 17 provides some personal preference design criteria to facilitate proper pump/trap system "breathing." A pump/trap has to "breathe in" motive steam and "breathe out" exhaust steam without significant restriction. Too small piping or piping loops in the motive or exhaust lines can severely effect operation. So too can a reduced port valve in either motive or exhaust line, or too small a balance tapping at the reboiler condensing source connection (shell tapping for shell-side steam, channel head tapping for tube-side steam) cause the system to perform poorly.

BALANCE & MOTIVE FOR 3X2 UNITS

PUMP/TRAP UNITS	BALANCE LINE (in)	MOTIVE LINE (in)
1	1	1
2	1.5	1.5
3	2	2
4	2.5	2
5	2.5	2.5

Δ **Figure 17.** Proper balance and motive line sizes are essential for effective pump/trap breathing.

Proper breathing means the motive steam must be without pipe or valve restriction or condensate build-up, and the exhaust steam must be able to freely balance without restriction or build-up. Improper line sizing, tapping location to the condensing source vapor tapping, or loop seal/reduced flow valve restrictions are some of the main causes that reviewed installed systems do not perform properly or at all. Failure to meet operational expectations is commonly not an equipment sizing issue, but rather a breathing issue. Consult the SPD2 manufacturer BEFORE ordering any new exchanger so that the proper vapor space tapping connection size and location can be provided to enable correct pump/trap installation.

Add a Spare SPD to Mitigate Interruptions

The previous recommendation of elevating the height of the reboiler outlet flange to 80 inches can reduce the amount of SPDs required to recover the condensate. However, that SPD reduction can be "reinvested" to mitigate process interruptions.

Consider that every mechanical product can require servicing at some time. So, if just the exact amount of SPDs is used to drain a reboiler, at the moment when one ceases to work properly it can be required to either discharge condensate to grade, operate on bypass (which can further pressurize the return header and lose some reboiler duty), or switch over reboilers. To mitigate such interruptions, install an extra SPD in the process design to operate as a functioning spare. That investment enables site operators to check the performance of each unit at regular intervals by opening a drain valve connected to a lower SPD tapping. Since it is less likely that more than one unit would require service at the same time, the SPD requiring attention can be repaired while the reboiler remains in operation.

A Vertical Thermosiphon Reboiler Review

Since much has been reviewed already regarding horizontal reboiler design, it can be interesting to review a vertical reboiler case history using shell-side steam, an ISC scheme, but with a level pot and outlet control valve acting as the steam trap. The installation background includes a condensate return header with "P4" back pressure of 1.8 barg, flash steam at 130°C, "P1" inlet pressure of 3.7 barg, and a highly over-surfaced reboiler.

The original design engineer must have felt that a P1 of 3.7 barg would be more than adequate to discharge into the expected P4 back pressure. However, the site encountered such severe water hammer, that for over 10 years the condensate was dumped to grade. There are three design scenarios to review:

- Large over-surfacing exposed to steam (original install)
- Small surface exposed to steam, but high residence time
- Large over-surfacing with pump/trap option.

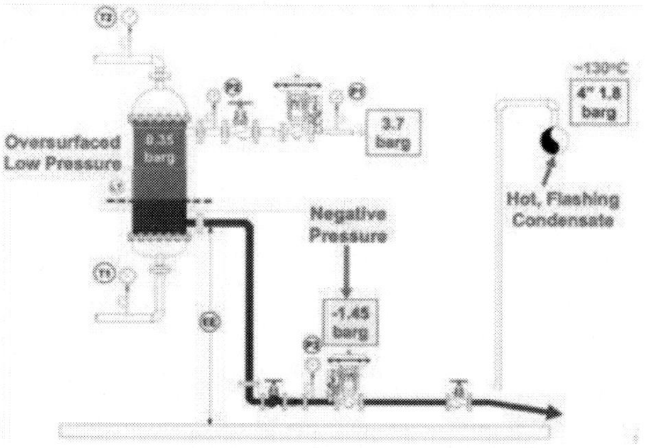

Δ **Figure 18.** Low steam pressure (0.35 barg) in reboiler has insufficient pressure to drain into condensate header (1.8 barg).

Figure 18 reveals the pressures of the first scenario. Although P1 (3.7 barg) was high, the large surface area caused P2 to drop drastically (0.35 barg). The pressure from P2 of

~0.35 barg (plus head) was incapable of discharging into the P4 back pressure of 1.8 barg, so the condensate was wasted.

The site decided to change the level transmitter that controlled the drain control valve to create a very high flood line (Figure 19). High flooding so reduced the steam condensing area that the P2 steam pressure elevated to 2.2 barg to supply the required heat to meet demand. This resulted in a positive pressure differential of 0.4 barg which was more than adequate to discharge condensate into the header. But, the reboiler was so over-surfaced that the high flood line resulted in significantly cooled condensate at the bottom of the reboiler, which subcooled further in the level pot so that the outlet temperature was approximately 92°C. Can it be imagined what happens when discharging colder non-flashing 92°C condensate into a condensate header full of flash steam at 130°C? The result was substantial collapse hammer so severe that the site had to again discharge the condensate to grade. Additionally, having the flood line so near the P2 steam inlet connection was also a source of instability and an undesirable aspect of this design.

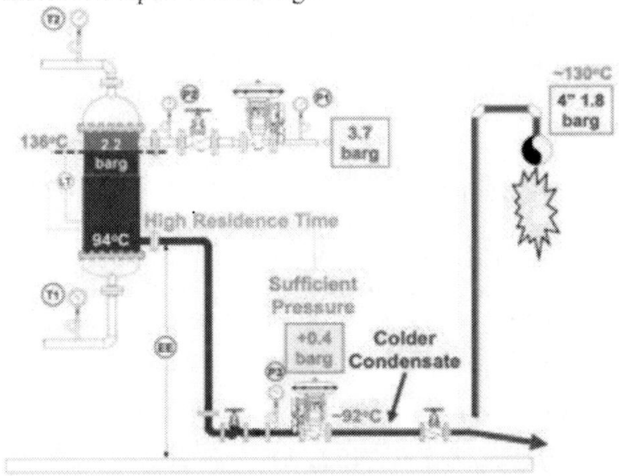

Δ **Figure 19.** High fooding elevates steam pressure (2.2 barg), residence drops temperature (92°C), creating hammer (130°C).

Figure 20 reflects the recommendation to install an SPD2 system. In this scenario, the steam space is returned to the original condition, allowing steam vapor to heat as much of the tube surface as possible. The P2 pressure drops to 0.35 barg and that pressure is equalized through the balance line to the reservoir pipe/tank. Condensate flows from the reboiler into the reservoir and pump/trap, but by itself cannot overcome the negative pressure differential at the onset, UNTIL the condensate level rises in the SPD2 and activates the 3.7 barg steam to be used as the motive source to pump a positive displacement of condensate (8 gallons/cycle) into the return header. Since the condensate is flashing at near 105 - 108°C, the flash can create a cushion while discharging into the flash steam header at 130°C. This system can handle the over-surfacing and stall condition without creating hammer or back-up into the reboiler.

In a scenario where the condensate might be significantly subcooled and non-flashing, it is possible that hammer could occur. There are several CONCEPT options for non-flashing liquid flow into the return header that could be considered with required evaluation and design by a knowledgeable

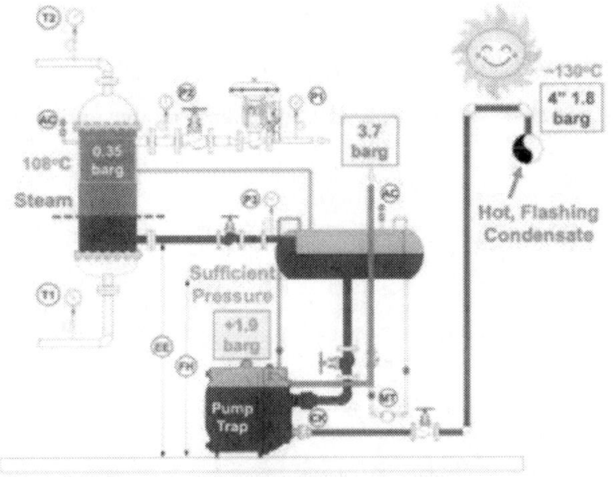

Δ **Figure 20.** Low reboiler steam pressure (0.35 barg) requires secondary pressure (3.7 barg) to discharge into condensate header.

engineer.

Those concept options could include discharging condensate into the side of the main header rather than the top where it might more readily collapse flash steam. There might be multiple side entry insertion points to reduce input build-up at any spot, use of globe valves to restrict flow velocity, incorporate check valves to prevent reverse flow, and inserting flow with angled entry rather than at right angle to facilitate integration of liquid-to-liquid without crossing into the flash steam flow.

In certain installations, a bottom feed concept arrangement might also be considered. However, all of these concept options could create severe collapse hammer in the condensate header, particularly if steam were to somehow be injected through the connecting line into the liquid condensate. Therefore, for safety requirements, such a concept design must be carefully evaluated, and reviewed-with appropriate hazop analysis performed-by a knowledgeable engineer to determine if suitable for any particular installation and must absolutely ensure there is no possibility of steam being injected through the connecting line into the condensate header liquid. As such, these concepts should never be used with OCC because the control valve may blow steam into the system and cause severe damage.

Some Development Background

My first work with the identifying and naming the stall phenomenon began in 1978 while reviewing a damaged heat exchanger as a member of the Philadelphia Chapter of Operating Engineers. After just one year of operation, the lower section of the tube bundle had so many leaks that it resembled Swiss cheese. This created a determination to understand why and how to mitigate the occurrence where condensate could not elevate out of an exchanger, resulting in such severe corrosion. The non-movement of condensate reminded me of the Piper Cherokee I trained/soloed on in 1975 where it was necessary to stall the aircraft and then safely pilot out of the situation, so this concept was used to name and explain the condition.

In a short period of time, it became evident that P2/P3 steam pressure could reduce below the P4 back pressure frequently as demand load decreased, and this phenomenon could be anticipated by charting the stall point graphically. On the first graph created it was noted:

- Positive pressure differential only requires a trap
- Negative pressure differential needs only a pump, but
- Varying differential requires a pump/trap combination.

Understanding the cause, it became possible to solve stall conditions by matching a separate SPD1 pump with an oversized float trap on the outlet (to account for the higher-than-average flow discharge rate). Later the dream of combining both mechanisms in one body was realized in a pump/trap SPD2 model referred to as "*The Original*" (Figure 11). Ultimately, to improve the evaluation process, the *Extended Stall Chart* used herein was published in 2004. Since that time, it has been used all over the world to anticipate and solve stall in over-surfaced exchangers by charting the stall point and incorporating pump/trap technology for applications with varying positive/negative differential pressures.

Acknowledgement:

Special thanks to Justin McFarland for creating several of the graphics contained herein, and to Henry Kister, Norman White, and Andrew Mohr for their kind review and comments.

Author Bio:

JAMES R. RISKO, CEM, PEM, MBA is president of TLV Corporation, Charlotte, NC, responsible for US and Canadian operations. He has 44 years' steam systems experience, authored more than 60 technical articles, provided webinars to over 2,500 attendees globally, and presented at IETC, AIChE, IPEIA, and AFPM conferences. He co-invented the world's first **combination pump/traps** and created the "***Extended Stall Chart***" for draining stalled heat exchangers. A past chairman of the FCI, he is currently active in FCI, ANSI, and ISO standards development. Jim is an avid tennis player and has three energy management certifications.

Email: Risko@TLVengineering.com
Cell: (704) 641-8959

LinkedIn: https://www.linkedin.com/in/jim-risko-aba7578

References:

1. **Kister, Henry Z.,** "Controlling Reboilers Heated by Condensing Steam or Vapor," *Chemical Engineering*, New York, NY, pp. 22 – 32, https://www.chemengonline.com/controlling-reboilers-heated-condensing-steam-vapor/ (July 2020)

2. **Risko, James R.**, "Steam Heat Exchangers are Underworked and Over-surfaced," *Chemical Engineering*, New York, NY, pp. 58–62, https://www.tlv.com/global/US/articles/steam-heat-exchangers-are-underworked-and-over-surfaced.html (Nov. 2004)

3. **Risko, James R.,** "Why Bad Things Happen to Good Steam Equipment," *Chemical Engineering*, New York, NY, pp. 50 – 58, https://www.tlv.com/global/US/articles/why-bad-things-happen-to-good-steam-equipment.html (Mar. 2015)

4. **Risko, James R.,** "Stop Knocking Your Condensate Return!," *Chemical Engineering Progress*, 112 (11), pp. 27–34, https://www.tlv.com/global/US/articles/stop-knocking-your-condensate-return.html (Nov. 2016)

5. **Risko, James R.,** "Steam Trap Management: Do Something; Anything. Please!," *Chemical Engineering Progress*, 113 (10), pp. 64–72, https://www.tlv.com/global/US/articles/steam-trap-management.html (Oct. 2017)

6. **Risko, James R.,** "Allocate New Plant Focus to Steam System Design-Part 2," *Hydrocarbon Processing, 98* (2), pp. 49–52. https://www.tlv.com/global/US/articles/plant-focus-on-steam-system-design-pt2.html (Feb. 2019)

619669 Troubleshooting Tunnel Vision: Packing Change Blamed for Premature Flooding

Abstract:

A process tower in a plant underwent a packing change to new 4th generation random packing as a way to gain needed capacity. The production team also used this time to do some other small maintenance on the tower and auxiliary equipment. Upon restart the tower experienced premature flooding at well below design rates. Everyone instantly blamed the new packing for this flooding even though the author insisted conditions were well below the hydraulic capacity of the packing. The tower was unpacked and repacked several times and finally the production team put the old packing back in; but still the problem persisted.

Looking into the other things that were changed during the turnaround it was discovered that a very small change had a tremendous and unexpected impact on the process. This was eventually linked to be the cause of the issue. The packing change had masked the real cause of the issue which extended the troubleshooting effort; costing more lost production.

This paper captures the summary of the troubleshooting investigation and the unique learnings that occurred during the investigation. It also serves as a reminder that not all root causes are obvious ones.

619539 Effect of Downcomer Design in Distillation at High Pressures

Abstract:

Distillation at higher pressures require careful considerations for the design of downcomers because the flooding mechanism is often governed by downcomer flood. Studies on trayed columns have shown that the mechanism of coalescence and separation of the two-phase liquid-vapor mixture in the downcomer is significantly more complicated as pressure increases. In terms of design of the downcomer, both the downcomer entrance area as well as the area available for the two-phase flow of froth within the downcomer are of importance. In this presentation, we describe research on downcomer performance conducted by FRI's legendary expert - Tak Yanagi. An analysis and interpretation of his findings on the effects of downcomer design in high-pressure distillation will be discussed.

617898 Debottlenecking of a Batch Distillation Process: Impact of Condenser Configuration

Abstract:

During the manufacture of a new product, a batch distillation step required considerably longer time than original anticipated, and it became a bottleneck for production. The distillation step was used to remove organic solvents from the target product in a column equipped with structured packing and a reflux condenser. A dynamic simulation model was used to represent a typical batch at the original operation conditions. The simulation results were then used to perform capacity rating calculations for the column and the condenser at different times during the patch. These calculations indicated that flooding in the tubes of the condenser was the key limiting factor. To overcome this limitation, the condenser configuration was changed, moving away from the reflux mode, that is, with vapor and condensate flowing in the same direction inside the tubes. After this change in configuration, the throughput limitation in the unit was removed, and a distillation time lower than 1/3 of the original time was consistently demonstrated in the plant. In addition, a higher distillate purity was achieved.

Sieve Tray Stability Factor Based on Efficiency Data

By

Daniel R. Summers, P.E.
Retired

Presented at the 2021 AIChE Spring Virtual National Meeting
Kister Distillation Symposium

Paper 79e
April 21, 2021

Abstract

The concept of Tray Stability was introduced by the author back in 2010[1]. The data used to determine this modified Froude number term was based on simulator data and the tray hydraulic parameter examined was weeping. This paper serves to validate this stability factor concept by demonstrating when tray efficiency data "falls off" as one turns down the capacity of a trayed distillation tower. That information is readily available for sieve trays from the publicly disclosed data that has been taken by Fractionation Research Incorporated (FRI) over the past 70 years[2]. This paper will look at that data to demonstrate there is a relationship between tray stability and tray efficiency and what physical properties and mechanical design features most affect that relationship.

Tray Stability

Tray Stability is defined as the square root of the dry tray pressure drop over the hydrostatic head on a distillation tray, see Equation 1 below. It is basically a Froude number and this hydraulic parameter was introduced to the public in 2010[1]. The dry tray pressure drop correlation for sieve trays was introduced in 2009[3] and updated in 2017[4] and the Hydrostatic Head of sieve trays was introduced by Charlie Colwell in 1981[5].

$$\eta = [\Delta P_{DRY}\, \rho_W / (\rho_L\, HS)]^{0.5} \qquad \text{(Eq. 1)}$$

Dry Tray Pressure Drop

The updated 2017 equation for dry tray pressure drop of sieve trays is as follows;

$$\Delta P_{DRY} = 12\;\; \rho_V\, (V_H/C_V)^2 / (2\, g_C\, \rho_W) \qquad \text{(Eq. 2)}$$
$$C_V = K\, (Dp/Pitch)^{0.10} \qquad \text{(Eq. 3)}$$
$$K = 0.997 - \{0.34 / [1.0 + (4.925\, TT/Dp)^{3.582}]\} \qquad \text{(Eq. 4)}$$

Where,

η = Stability Factor
Dp = Hole Diameter, inches
TT = Tray Thickness, inches
Pitch = Center to center distance between holes, inches
ΔP_{DRY} = dry tray pressure drop, inches of Water
V_H = Ub/fp, Hole Velocity, ft/sec
ρ_V = Vapor Density, lb/ft^3
C_V = Hole (Orifice) discharge Coefficient
ρ_W = Water Density = 62.428 lb/ft^3

Hydrostatic Head

Colwell's hydrostatic head correlation is as follows;

$$HS = (\alpha\ How) + K_0\ [(\alpha^{0.5}/Cd)\ (GPM/Bw)]^{2/3} \quad \text{(Eq. 5)}$$

Where,

$\alpha = 1/(1+\varepsilon)$ — Relative Froth Density

$\varepsilon = 12.6\ Fr^{0.4}\ fp^{-0.25}$

$Fr = 12.0\ (\rho_V/(\rho_L-\rho_V))\ (Ub^2/(g_C\ HS))$ — Vapor Froud Number

ρ_L = liquid density, lb/ft^3

g_C = gravity constant, 32.174 ft/sec^2

HS = Hydrostatic Head, inches of liquid

$Ub = V / (3600\ \rho_V\ Ab)$, ft/sec

V = Vapor Rate, lb/hr

Ab = Tray Active Area, ft^2

How = Outlet Weir Height, inches

Cd = Weir Coefficient

$Cd = 0.61 + (0.08\ K_1)$ for K_1 less than or equal to 8.135

$Cd = 1.06\ (1 + (1/K_1))^{1.5}$ for K_1 greater than 8.135

Cd = 1.06 for How = 0.0 inches

K_1 = (HF – How) / How

$HF = HS / \alpha$ — Froth Height, inches

GPM = Liquid Rate, gpm

Bw = Weir Length, inches

fp = Open Area/Ab — Fraction Open Area

K_0 = 0.527 for round sieve openings

Discussion

The (public domain) FRI Sieve tray data was examined to see if it substantiates the stability work performed by the author back in 2010. Keep in mind that the author had already looked at the public domain FRI tray data in 2018[6] to find a relationship between stability and tray efficiency for Bubble Cap type trays. For the current paper however, the data examined was established by Fractionation Research Inc. (FRI) in Alhambra, CA between the years 1957 and 1980. The data was taken in 2 industrial size towers (48 inch and 96 inch diameter) at total reflux and shows the tray efficiency vs. vapor load. There were more than 100 data sets taken during this time period with numerous compounds and pressures operating from deep vacuum to 500 psia. Figure 1 shows an example of one such data set with cyclohexane/n-heptane operating at 24 psia pressure. One can see that the tray efficiency drops from about 80% to less than 50% as the C_B falls below 0.1 ft/s. This could be interpreted as an indication of minimum efficient operation.

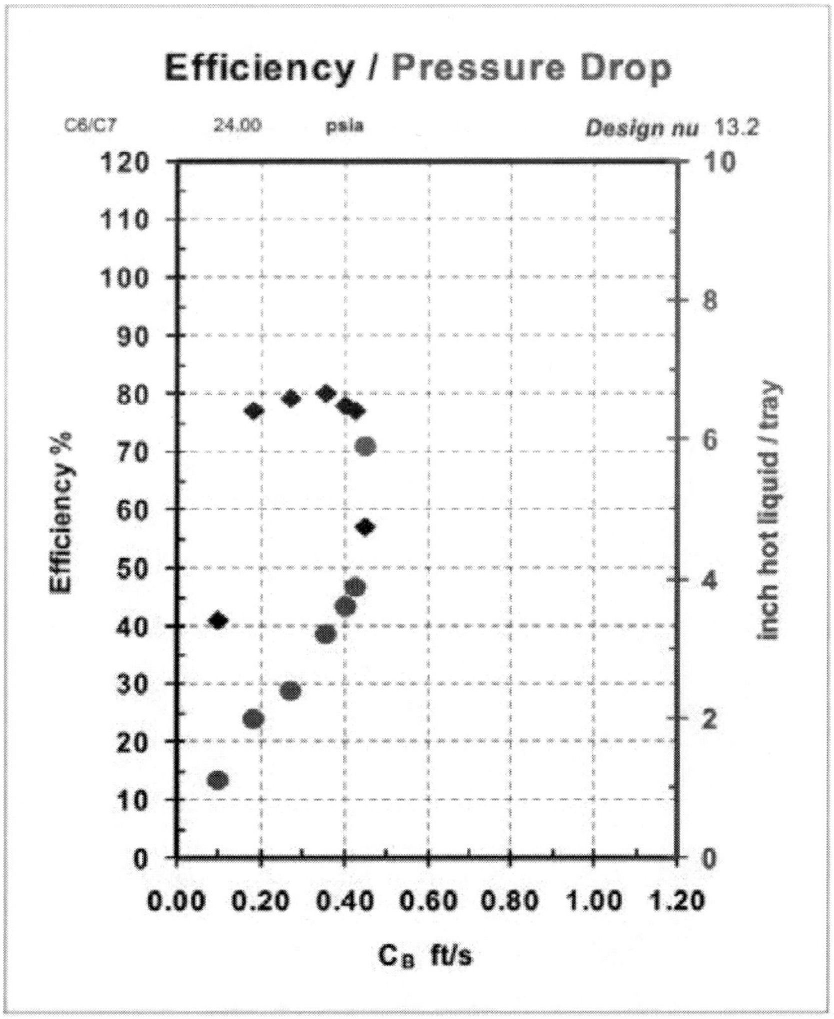

Figure 1 – FRI Run Numbers 1323 to 1329, January 1957

Through the 1543 pages of available FRI data, the author looked for turndown data such as what is shown in Figure 1. The black diamonds in this figure represent tray efficiency and the blue circles represent tray pressure drop. The author also looked for data that showed no signs of weeping. The data sets from FRI included an indication, in the data text, if weeping was observed from below the trays. Recent work at FRI has shown that even minor wall weeping/leakage can cause a significant loss of tray efficiency at low vapor rates. Therefore, the author decided to avoid any data that contained observations of weeping in the data. The author also avoided all 2-Pass tray data because 2-Pass trays can exhibit their own instabilities as discussed in the 2010 paper. In addition, only data with the perforation punched downwards was employed in this study.

The non-weeping data sets were then analyzed with the above Stability equations and a plot of tray stability was added to each set. This resulted in a number of plots that look like the example shown below in Figure 2. Since the calculated Stability line appears to be linear to the vapor load C_B, it was then concluded that tray efficiency can be easily be plotted against the calculated Stability factor with no degradation in the shape of the curve, see Figure 3. Now, one can visually determine the tray stability factor at turndown where tray efficiency starts to fall and the minimum stability factor can be determined where good tray efficiency can be maintained.

The author was able to identify a total of 91 data sets that exhibited sufficient credible evidence of efficiency reduction at turndown, see attached Appendix A. Appendix B is also provided to show the specific calculations entailed for the stability determination shown in Figure 2. Please note that the Colwell clear liquid height calculation (HS) is nearly constant at about 1.6 inches of liquid for the total reflux data runs 14115 to 14119 in Appendix B.

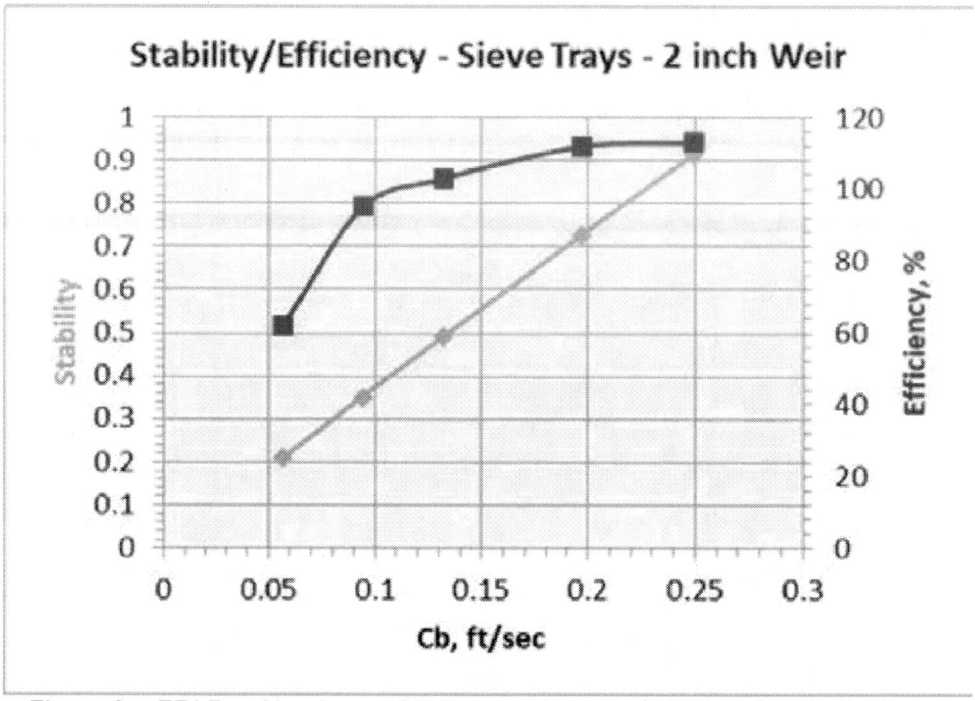

Figure 2 – FRI Run Numbers 14115 to 14119, with Calculated Stability Added

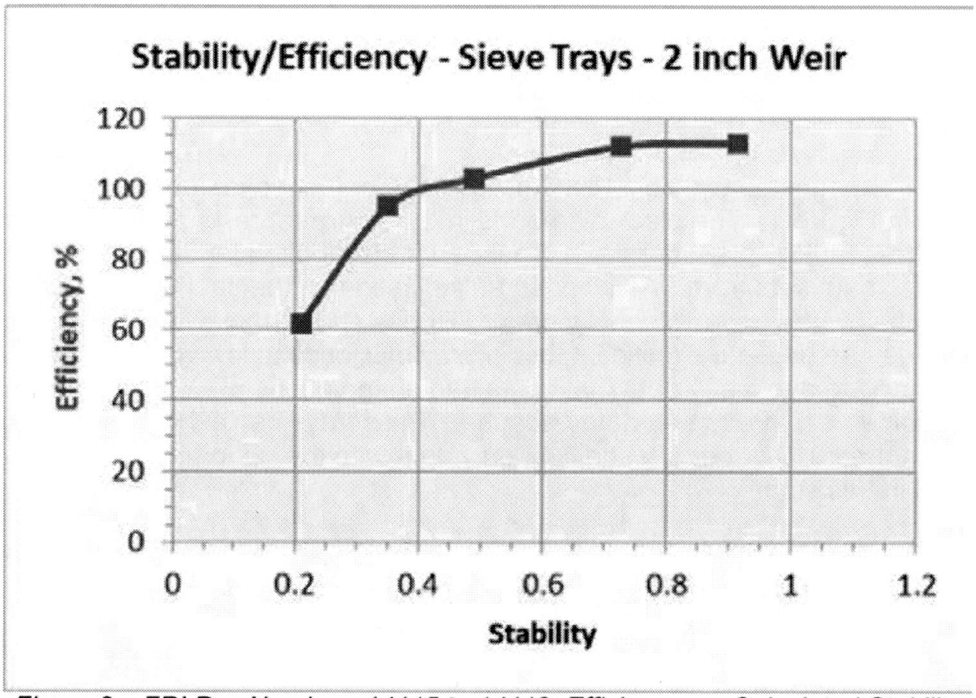

Figure 3 – FRI Run Numbers 14115 to 14119, Efficiency vs. Calculated Stability

For each of these data sets, a determination had to be made as to what exactly was the minimum stability factor. This can be very subjective since the difference between a good tray efficiency point and a "bad" tray efficiency point can have a large variation in stability factor value. The data taken by FRI back in the 1950's and 1960's was primarily focused on maximum capacity determination and was not as focused on finding exactly where tray efficiency drops off at turndown. The data set in the example shown as Figure 3 above shows good efficiency at a stability factor of 0.35 and much poorer efficiency at a stability factor of roughly 0.2. Please note that between stability factors of 0.35 and 0.9, the tray efficiency in this figure is fairly constant between 95 and 110%. It is very difficult to determine exactly from Figure 3 at what low stability factor the tray efficiency would have started to fall significantly below approximately 90%. Therefore, the data in the attached Appendix A contains the best "subjective" guess as to what the minimum stability factor should be and a range where that minimum stability factor will be found. This was performed for all 91 data sets.

Appendix A shows the data from FRI with the subjective stability factor determinations. This appendix also contains a number of key mechanical as well as physical property parameters for each analyzed set. The mechanical parameters include; outlet weir height, outlet weir length, tray spacing, sieve hole diameter, hole area, tray thickness, and type downcomer. The physical property variations are; type system, operating pressure, surface tension, weir loading, vapor density and liquid density.

One of the first things noticed by the author when analyzing the data in Appendix A, was that there is a strong influence of physical properties on the minimum stability factor. The data for ½" diameter holes, 2" tall outlet weirs, 24 inch tray spacings, 16 gauge tray decks, 33 to 37 inch weir lengths and 8.3 to 8.6% open area, is shown below as Figure 4. This curve uses surface tension as the correlating parameter. The data shown in Figure 4 result in a well-defined curve. The best-fit equation for the curve through this data is shown in Figure 4.

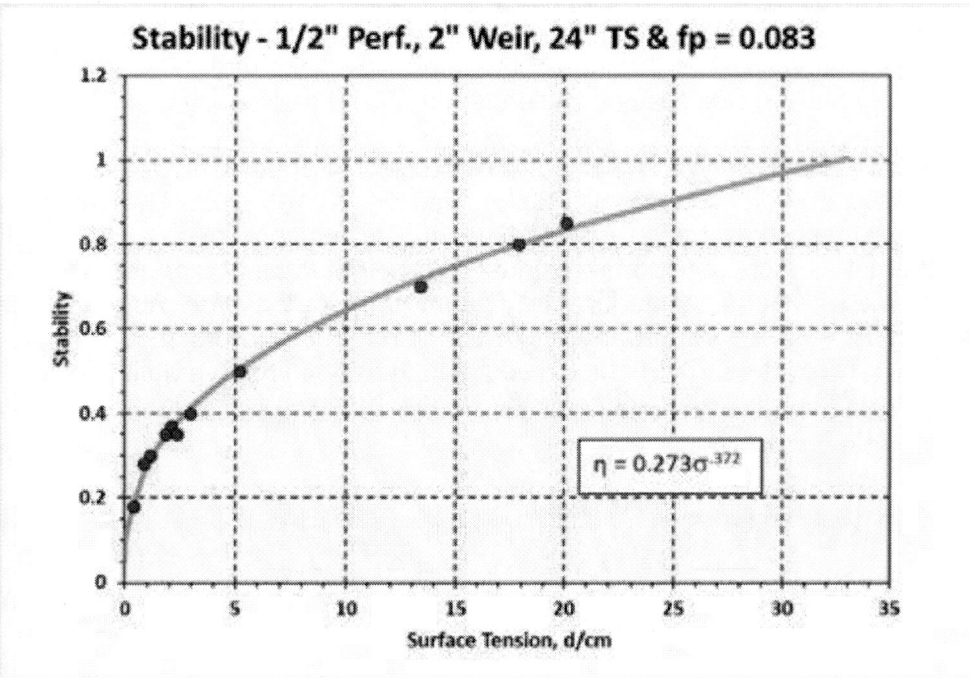

Figure 4 – Stability at Constant Geometry

Other physical properties such as liquid density, vapor density and delta density can be used as potential correlating parameters as well. The best least squares fit for the data shown in Figure 4 was found by using vapor density as shown here in Figure 5. Using vapor density, in the author's opinion, made the most sense because it easily is indicative of operating pressure, but it is also one of the key physical properties in determining the stability calculation itself.

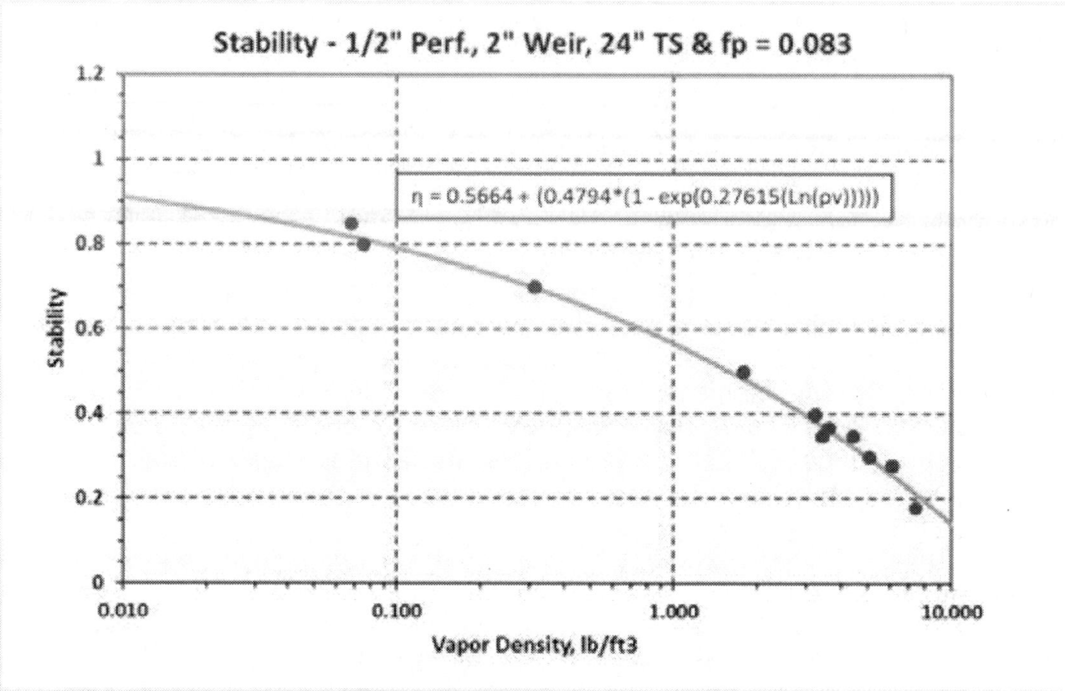

Figure 5 – Stability at Constant Geometry

After establishing that vapor density has the best fit and it has a big influence on tray stability, several mechanical features, and their effect, were examined next. The data in Appendix A has quite an assortment of tray mechanical design features that may need to be included into the equation shown in Figure 5. The mechanical features that were examined are outlet weir height, percent open area, weir loading, hole diameter, and tray spacing.

Outlet weir height's effect on stability was examined first. It appears to be a small (but significant) effect when you look at Figure 6, for example, at 165 psia. The outlet weir height effect should have been accounted for already in the Colwell correlation above, see Equation 5. However, in the author's opinion, some residual weir height influence appears to be present. There was no data with perfectly identical tray geometries for the wide range of outlet weir heights examined in Appendix A (from 0.0 to 4.0 inches), but Figure 6 shows that the affect is there regardless. The data shows a slight reduction in the minimum stability factor at taller outlet weirs. The adjustment was added to the equation shown in Figure 5 and is shown here as Equation 6.

$$\eta_{MIN} = [0.5664 + \{0.4794\,(1 - e^{(0.27615(Ln(\rho v)))})\}]\,\{1.1 - (0.05\,H_{OW})\} \qquad \textbf{(Eq. 6)}$$

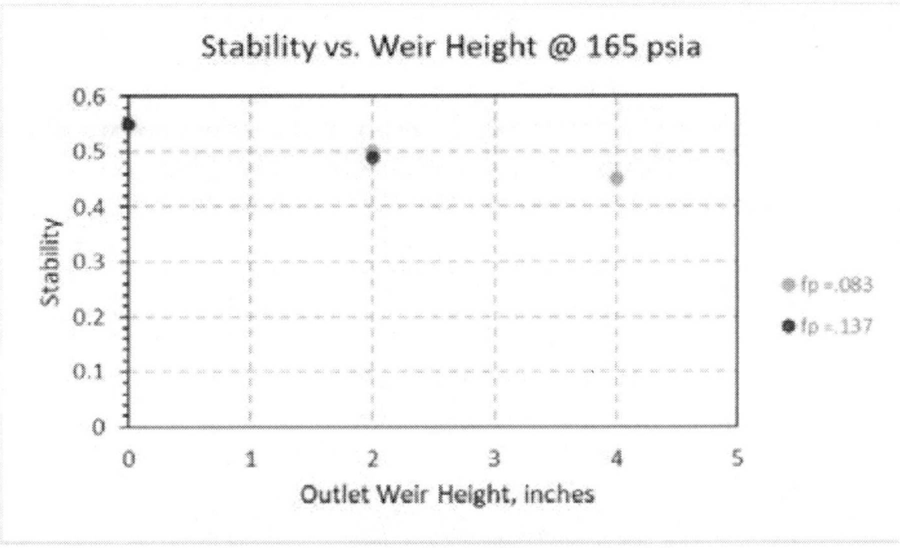

Figure 6 – Outlet Weir Effect on Stability

Open area was examined next. The open area effect in Figure 7 below, is shown as "fp" which represents fraction perforation or percent open area divided by 100. It is clear that open area has an effect on minimum tray stability factor. The Appendix A data was grouped into three major open area categories. One can see that lower open areas require higher minimum stability factors and higher open areas reduce the minimum stability factor.

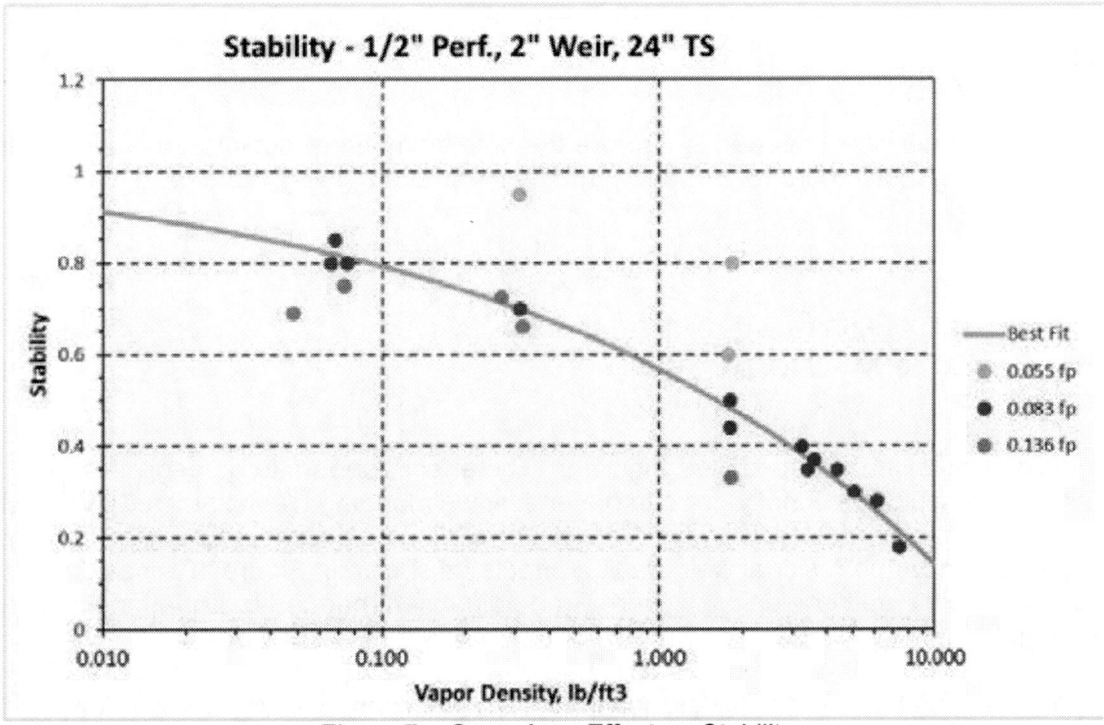

Figure 7 – Open Area Effect on Stability

The adjustment (multiplier) for the open area effect was added to Equation 6 and is shown here in Equation 7 shown below.

$$\eta_{MIN} = [0.5664 + \{0.4794\ (1 - e^{(0.27615(Ln(\rho V)))})\}]\ \{1.1 - (0.05\ H_{OW})\}\ (0.083/fp)^{0.33} \qquad \text{(Eq. 7)}$$

There also appears to be a small hole diameter effect as can be seen here in Figure 8. This plot was generated at constant open area, weir height and operating pressure. The hole diameter affect appears to be linear and significant.

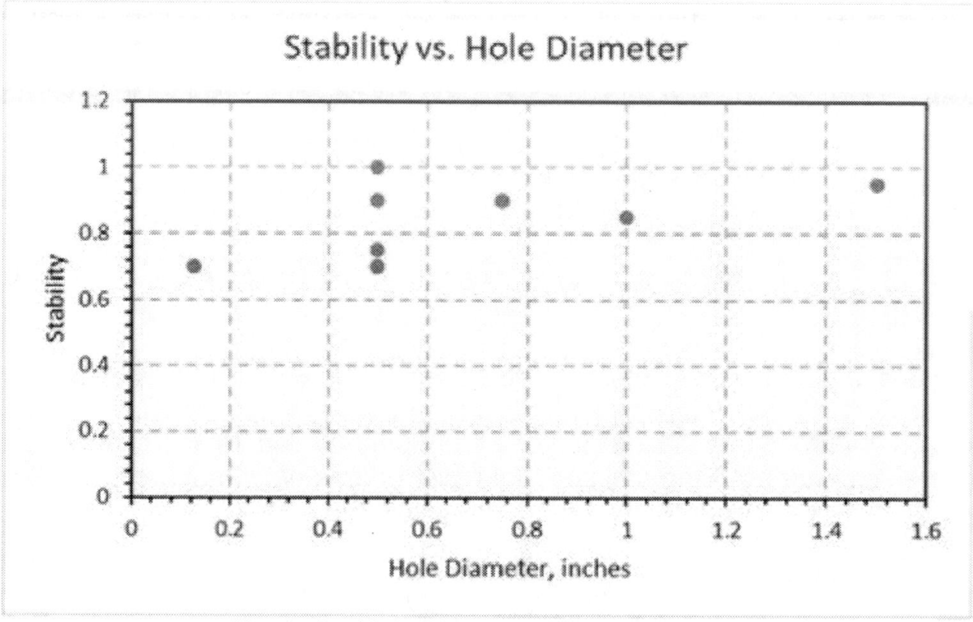

Figure 8 – Hole Diameter Effect on Stability

The adjustment (multiplier) needed to capture the hole diameter effect into equation 7 was added and is now shown here as Equation 8. The minimum stability factor equation is getting quite lengthy.

$$\begin{aligned}\eta_{MIN} = \ & [0.5664 + \{0.4794\ (1 - e^{(0.27615(Ln(\rho V)))})\}] \cdot \\ & \{1.1 - (0.05\ H_{OW})\}\ (0.083/fp)^{0.33} \cdot \\ & [0.858 + 0.142(Dp/0.5)] \qquad \text{(Eq. 8)}\end{aligned}$$

Tray spacing was examined next, see Figure 9. The effect appears to be negligible as, theoretically, it should be. Finally, for completeness, weir loading was examined, which should already have been incorporated into the Colwell correlation as shown in Equation 5 above. The data appears to have no residual weir loading effects as shown in Figure 10 since the data is too scattered. In the author's opinion, there is no adjustment needed to Equation 8 with regards to weir loading.

Therefore, equation 8 is the definitive equation for the <u>Minimum Stability</u> of Sieve trays based on FRI public domain data. A Parity plot for this equation was generated, shown as Figure 11, to see how well Equation 8 compares to all the data in Appendix A.

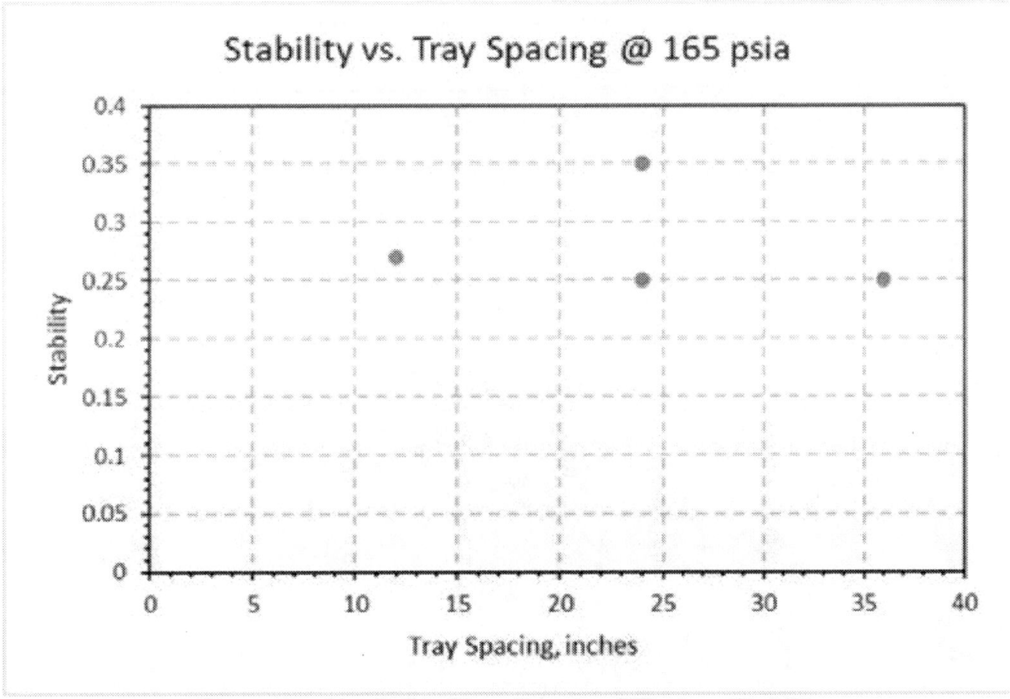

Figure 9 – Negligible Tray Spacing Effect on Stability

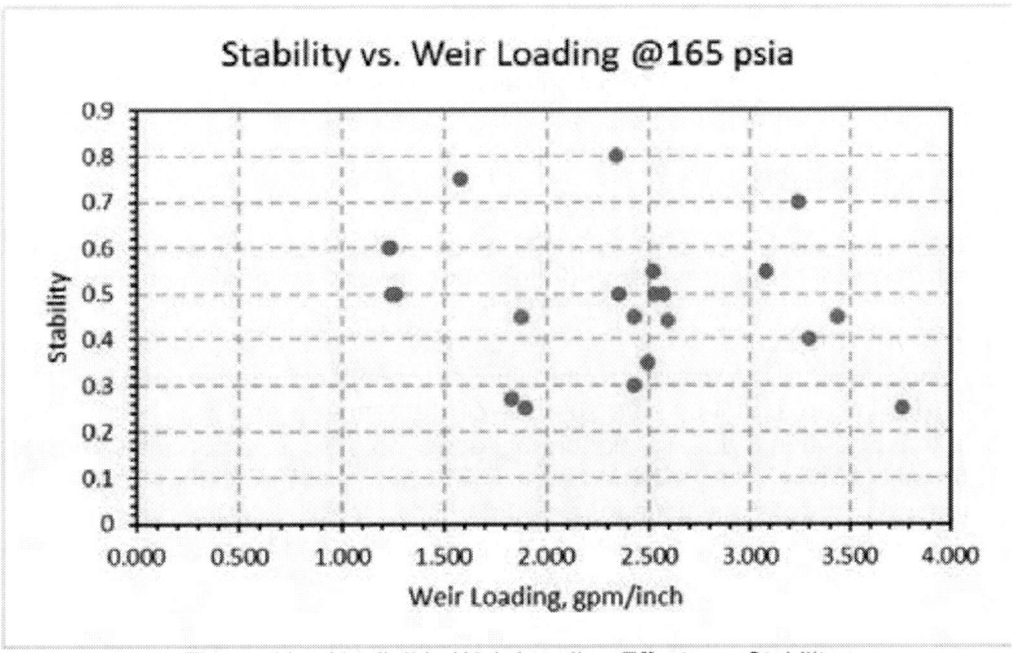

Figure 10 – Negligible Weir Loading Effects on Stability

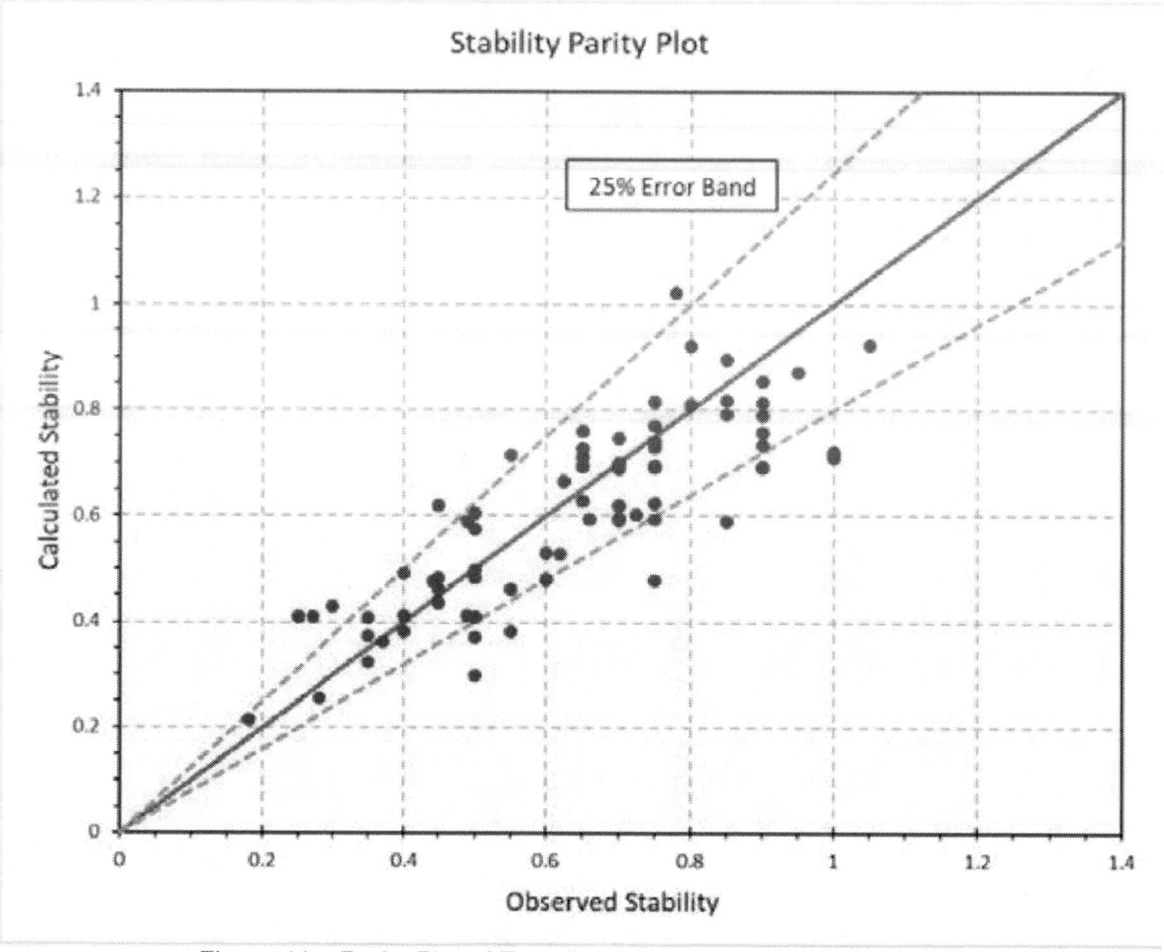

Figure 11 – Parity Plot of Equation 8 including a 25% Error Band

<u>Observations and Opinion</u>

Equation 8 shows us that the 2010 work by the author was limited in scope since the data he evaluated was based only on air-water and air-isopar simulator weeping observations. His conclusion in 2010 was that a fixed value of 0.6 was appropriate as a minimum guideline for the Stability Factor. With the current work here, which now includes real column efficiency data, the author now concludes that the minimum Stability Factor is heavily influenced by the operating pressure and to a limited extent by hole size, percent open area and weir height. Figure 11 above shows that Equation 8 does a reasonably good job of representing the data. However, because of the subjectivity of determining exactly where the tray efficiency "drops off" in the data sets examined, there is a significant amount of scatter built into the data shown in Appendix A. The author would have liked a narrower error band than 25% in the Figure 11 parity plot.

It is the author's opinion that Equation 8 will provide users another way to look at sieve tray turndown as compared to applying traditional weeping predictive methods that have led to poor results. It is also the author's opinion that similar to the sieve tray results shown here, fixed opening devices should also have a minimum stability curve similar to Equation 8.

Conclusions

An equation has been generated that defines the minimum tray stability factor value for Sieve Trays. The applicability of this equation covers a wide range of physical and mechanical parameters as shown here in Table 1. The equation has been set up to be extrapolatable outside of the experience bounds established in Table 1. This equation expands on the 2010 work by the author which was totally based on air-water and air-isopar simulator data.

Table 1 – Stability Factor Equation 8 - Bounds

Minimum	Parameter	Maximum	Units
12	Tray Spacing	36	inches
0	Outlet Weir Height	4	inches
3	Pressure	500	psia
0.24	Surface Tension	27	dynes/cm
0.004	Vapor Density	8	lb/ft3
20	Liquid Density	53	lb/ft3
12.6	Delta Density	53	lb/ft3
0.125	Hole Diameter	2.625	inches
0.2	Weir Loading	3.85	gpm/inch
4.7	% Open Area	17.4	--
0.0625	Tray Thickness	0.25	inches

In comparison to the author's 2010 work, the new minimum stability factor can now have values less than the 0.6 fixed value (as given in 2010) when column operating pressures are high. However, when operating pressures are low, stability factor values will more than likely need to be higher than the original 0.6 value.

Alternative equation

An alternative equation based on Surface Tension for the minimum Stability Factor can be used, see Equation 9 below. The parity plot for this equation, Figure 12, is really no different than the Parity plot for Equation 8. Therefore, for those people who think that the controlling physical property to best represent the pressure effect is surface tension, then Equation 9 is a satisfactory alternative.

$$\eta_{MIN} = 0.273\ \sigma^{0.372}\ \{1.1 - (0.05\ H_{OW})\}\ (0.083/fp)^{0.33}\ [0.858 + 0.142(Dp/0.5)] \qquad \text{(Eq. 9)}$$

Acknowledgement

The author would like to acknowledge the help he received years ago from Zane King of Sulzer. As an intern, he was able to take the public domain data from FRI and plot it into a usable and visual form for the author so that (after he retired from Sulzer) his work was made so much easier. Also, the author would like to acknowledge FRI for 70 years of industrial distillation data. FRI is a not-for-profit Worldwide Consortium of approximately 85 companies involved with distillation design ranging from equipment manufacturers to engineering companies to operating companies. FRI currently located in Stillwater, OK, is unquestionably the world leader in distillation technology. FRI has 3 important deliverables to the Distillation Industry;

1.) Industrial Sized Column Operating Data
2.) Design and Rating Correlations for both Packings and Trays
3.) Design Practices Handbook.

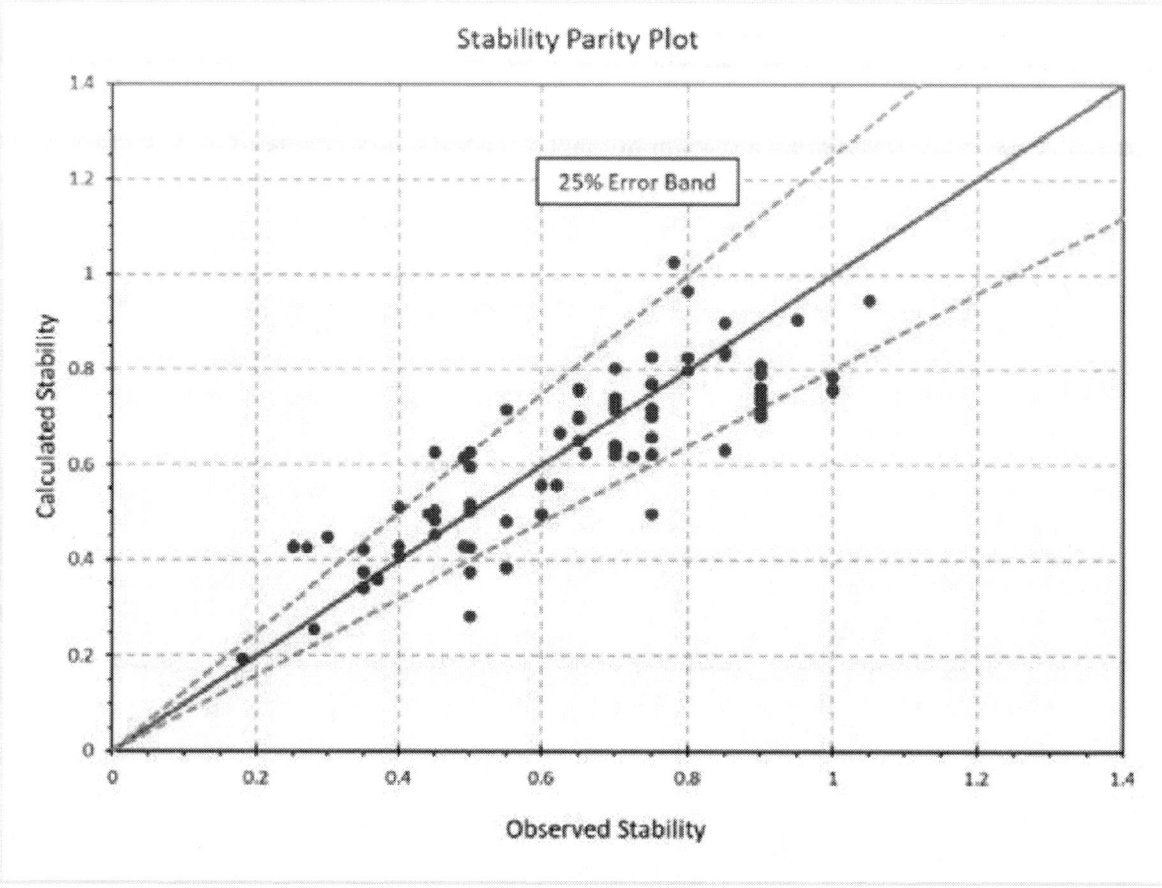

Figure 12 – Parity Plot of Equation 9 including a 25% Error Band

References

1.) Summers, D. R., Spiegel, L., and Kolesnikov, E., "Tray Stability at Low Vapor Load", Conference Proceedings of "Distillation and Absorption 2010", p.611, Eindhoven, The Netherlands, September 12-15, 2010.
2.) FRI Sieve Tray Data Base – Public Domain files, Oklahoma State University Special Collections and University Archives
3.) Summers, D.R., "Dry Tray Pressure Drop of Sieve Trays", Chemical Engineering, June 2009, pp 36-39.
4.) Summers, D.R., Cai, T. J. "Dry Tray Pressure Drop of Sieve Trays Revisited", Chemical Engineering, August 2017, pp 38-41.
5.) C. J. Colwell, "Clear Liquid Height and Froth Density on Sieve Trays", Ind. Eng. Process Design and Development, 1981, Volume 20, pp 298-307
6.) Summers, D.R., "Bubble-Cap Tray Vapor Turndown", Chemical Engineering, February 2018, pp 38-41.

Appendix A – Sieve Tray Stability Data Rendered from Efficiency Plots

Data Set	Weir Height, in.	Weir length, in.	TS, in.	Hole Dia, in.	Ab, ft2	Ah, ft2	fp	Type DC	Press, psia	System	Stability	Error Band, +/-	Notes	Surface Tension, d/cm	Vapor Density, lb/ft3	Liquid Density, lb/ft3	Tray Thickness, in.	Weir Loading, gpm/inch
1428-1435	2	37	24	0.5	9.25	0.77	0.0832	Straight	4	C6/C7	0.85	0.150		20.08	0.0675	44.7	0.0625	0.484
1451-1456	2	37	24	0.1875	9.25	1.35	0.1459	Straight	24	C6/C7	0.62	0.050		14.38	0.3098	41.3	0.0625	1.811
1459-1465	2	37	24	0.1875	9.25	1.35	0.1459	Straight	4	C6/C7	0.7	0.100		20.16	0.0647	44.7	0.0625	0.784
1475-1480	2	37	24	1	9.25	1.35	0.1459	Straight	165	IC4/NC4	0.45	0.050		5.36	1.7400	30.9	0.0625	3.432
1527-1533	2	52	24	0.5	8.85	1.24	0.1401	Straight	4	C6/C7	0.9	0.300	SBW	20.16	0.0647	44.7	0.0625	0.321
1551-1559	2	52	24	0.5	8.85	0.75	0.0847	Straight	165	IC4/NC4	0.45	0.300	SBW	5.31	1.7600	30.8	0.0625	1.875
1571-1580	2	52	24	0.5	8.85	0.75	0.0847	Straight	24	C6/C7	0.75	0.150	SBW	13.6	0.3193	40.2	0.0625	0.700
1589-1593	2	52	24	0.5	8.85	0.75	0.0847	Straight	50	C6/C7	1.6	0.349	SBW (unclear eff drop)	10.58	0.6176	38.2	0.0625	1.667
1604-1618	2	52	24	0.5	8.85	0.75	0.0847	Straight	24	C6/C7	0.9	0.050	SBW	13.54	0.3174	40.2	0.0625	1.173
1620-1625	2	52	24	0.5	8.85	0.75	0.0847	Straight	4	C6/C7	0.9	0.100	SBW	18.89	0.0665	43.3	0.0625	0.288
1648-1654	2	52	24	1	8.85	1.31	0.1480	Straight	165	IC4/NC4	0.55	0.020	SBW	5.4	1.7300	30.9	0.0625	2.519
1707-1714	2	37	24	1	8.05	0.79	0.0981	Straight	24	C6/C7	0.85	0.050	Blanking	14.27	0.3073	41.1	0.0625	1.022
1806-1814	4	37	24	0.5	9.25	0.77	0.0832	Straight	24	C6/C7	0.65	0.050		13.79	0.3124	40.4	0.0625	1.065
1837-1844	4	37	24	0.5	9.25	0.77	0.0832	Straight	165	IC4/NC4	0.45	0.030		5.21	1.7800	30.8	0.0625	2.424
1848-1854	2	37	24	0.5	9.25	0.8	0.0865	Straight	24	C6/C7	0.7	0.100	Thick Trays	14.48	0.3122	41.4	0.25	1.351
1865-1870	2	37	24	0.5	9.25	0.8	0.0865	Straight	4	C6/C7	0.8	0.100	Thick Trays	20.32	0.0656	45	0.25	0.570
1874-1881	2	37	24	0.5	9.25	0.8	0.0865	Straight	165	IC4/NC4	0.441	0.020	Thick Trays	5.19	1.7900	30.7	0.25	2.589
1885-1892	2	37	24	2.625	9.25	1.39	0.1503	Straight	24	C6/C7	0.8	0.150	Bcap with no caps	14.25	0.3124	41.1	0.0625	1.554
1988-1996	2	37	24	0.5	9.25	0.462	0.0499	Straight	165	IC4/NC4	0.6	0.050		5.21	1.7600	30.8	0.0625	1.265
2000-2007	2	37	24	1	9.25	0.79	0.0854	Straight	24	C6/C7	0.85	0.200		14.66	0.3052	41.6	0.0625	1.238
2018-2025	2	37	24	1	9.25	0.79	0.0854	Straight	4	C6/C7	1.05	0.200		20.3	0.0705	45.2	0.0625	0.578
2057-2064	2	33.5	24	0.5	9.25	0.77	0.0832	Straight	4	C6/C7	0.8	0.100	Splash Baffle	17.91	0.0748	42.4	0.0625	0.507
2068-2074	2	33.5	24	0.5	9.25	0.77	0.0832	Straight	165	IC4/NC4	0.5	0.050		5.21	1.7800	30.7	0.0625	2.522
2107-2114	0	37	24	0.5	9.25	0.77	0.0832	Stepped	4	C6/C7	1.6	0.200		18.89	0.0665	43.2	0.0625	0.508
2398-2408	2	37	36	0.5	9.25	0.77	0.0832	Stepped	4	C6/C7	0.75	0.100		19.72	0.0686	44.3	0.0625	0.370
2458-2466	2	37	36	0.5	9.25	1.27	0.1373	Stepped	24	C6/C7	0.49	0.050		13.93	0.3160	40.8	0.0625	1.049
2489-2497	2	37	36	0.5	9.25	1.27	0.1373	Stepped	165	IC4/NC4	0.25	0.050		5.15	1.7800	30.7	0.0625	1.897
2501-2509	2	37	36	0.5	9.25	1.27	0.1373	Stepped	14.7	O-P XYLENE	0.75	0.100		16.69	0.2020	47.3	0.0625	1.159
2521-2528	2	37	36	0.5	9.25	1.27	0.1373	Stepped	0.773	O-P XYLENE	0.9	0.200		24.72	0.0155	51.9	0.0625	0.289
2643-2650	2	37	12	0.5	9.25	1.27	0.1373	Stepped	24	C6/C7	0.7	0.150		15.25	0.2957	42.2	0.0625	1.157
2656-2663	2	37	12	0.5	9.25	1.27	0.1373	Stepped	4	C6/C7	0.75	0.200		21.09	0.0644	45.8	0.0625	0.741
2670-2677	2	37	12	0.5	9.25	1.27	0.1373	Stepped	165	IC4/NC4	0.27	0.050		5.15	1.7900	30.7	0.0625	1.832
2829-2837	2	37	24	0.125	9.25	0.793	0.0857	Stepped	24	C6/C7	0.7	0.150		13.86	0.3029	40.5	0.0625	1.324
2863-2870	2	37	24	0.125	9.25	0.793	0.0857	Stepped	165	IC4/NC4	0.3	0.100		5.32	1.7600	30.8	0.0625	2.424
2873-2880	2	37	24	0.75	9.25	0.81	0.0876	Stepped	24	C6/C7	0.9	0.200		13.59	0.3130	40.2	0.0625	1.297
3111-3118	2	37	24	1.5	9.25	0.83	0.0897	Stepped	24	C6/C7	0.95	0.300		13.76	0.3177	40.4	0.0625	1.176
3130-3137	2	37	24	1.5	9.25	0.83	0.0897	Stepped	4	C6/C7	0.78	0.200		19.28	0.0691	43.8	0.0625	0.392
3149-3156	2	37	24	1.5	9.25	0.83	0.0897	Stepped	165	IC4/NC4	0.5	0.100		5.11	1.7900	30.7	0.0625	2.568
3206-3213	2	37	24	0.5	7.25	0.484	0.0668	Stepped	165	IC4/NC4	0.8	0.200	Blanking	5.09	1.8100	30.5	0.0625	2.341
3226-3233	2	37	24	0.5	7.56	0.785	0.1038	Stepped	24	C6/C7	0.8	0.200	Blanking	13.92	0.3276	40.8	0.0625	1.297
3264-3271	2	37	24	0.5	7.56	0.785	0.1038	Stepped	165	IC4/NC4	0.5	0.060	Blanking	5.15	1.8000	30.7	0.0625	2.351
3283-3290	2	37	24	1	7.4	0.817	0.1104	Stepped	24	C6/C7	0.8	0.200	Blanking	13.91	0.3213	40.6	0.0625	1.284
3302-3309	2	37	24	1	7.4	0.817	0.1104	Stepped	4	C6/C7	1.05	0.250	Blanking	19.39	0.0716	44	0.0625	0.581
3321-3328	2	37	24	1	7.4	0.817	0.1104	Stepped	165	IC4/NC4	0.7	0.100	Blanking	5.13	1.8100	30.6	0.0625	3.243
3359-3366	2	37	24	0.5	8.77	0.67	0.0764	Stepped	24	C6/C7	0.8	0.150	Blanking	13.91	0.3213	40.7	0.0625	1.151
3563-3570	0	37	24	0.5	9.25	1.27	0.1373	Stepped	165	IC4/NC4	0.5	0.100		5.12	1.8000	30.7	0.0625	3.081
3627-3634	0	37	24	0.5	9.25	1.27	0.1373	Stepped	24	C6/C7	1.1	0.100		14.16	0.3160	40.6	0.0625	1.341
3666-3673	4	37	24	0.5	9.25	1.27	0.1373	Stepped	24	C6/C7	0.6	0.050		14.1	0.3204	40.9	0.0625	1.778
3684-3691	4	37	24	0.5	9.25	1.27	0.1373	Stepped	4	C6/C7	0.45	0.050		19.39	0.0716	44	0.0625	0.541
4150-4156	2	30	24	0.5	11.2	0.883	0.0788	Straight	165	IC4/NC4	0.4	0.150		5.16	1.7900	30.7	0.0625	3.300
4199-4205	2	30	24	0.5	11.2	1.51	0.1348	Straight	24	C6/C7	0.66	0.100		14.23	0.3177	41.1	0.0625	2.180

4217-4223	2	30	24	0.5	11.2	1.51	0.1348	Straight	4	C6/C7	0.75	0.150		19.33	0.0730	43.9	0.0625	1.013
4235-4241	2	30	24	0.5	11.2	1.51	0.1348	Straight	165	IC4/NC4	0.333	0.100		5.16	1.8000	30.7	0.0625	3.767
4301-4307	2	37	24	0.5	9.25	1.27	0.1373	Straight	20	IC8/TOL	0.725	0.100		14.07	0.2690	41.8	0.0625	1.416
4318-4325	2	37	24	0.5	9.25	1.27	0.1373	Straight	3	IC8/TOL	0.69	0.050		19.28	0.0480	44.8	0.0625	0.462
4619-4625	2	40.75	24	0.5	7.86	0.57	0.0725	Straight	24	C6/C7	0.65	0.050		13.74	0.3230	40.5	0.0625	0.564
7001-7007	2	37	24	0.5	9.25	0.77	0.0832	Straight	300	IC4/NC4	0.37	0.050		2.1	3.6000	26.5	0.0625	2.838
7037-7042	2	37	24	0.5	9.25	0.77	0.0832	Straight	300	IC4/NC4	0.35	0.050		2.34	3.4200	26.9	0.0625	2.676
7046-7050	2	37	24	0.5	9.25	0.77	0.0832	Straight	400	IC4/NC4	0.3	0.050	Unclear Efficiency Drop	1.16	5.0300	24.3	0.0625	2.132
7055-7059	2	37	24	0.5	9.25	0.77	0.0832	Straight	500	IC4/NC4	0.18	0.020		0.4	7.3600	21.2	0.0625	2.335
7101-7108	2	37	24	0.5	9.25	0.77	0.0832	Straight	300	C3=/C3	0.4	0.050		2.97	3.2400	27.3	0.0625	2.454
7112-7119	2	37	24	0.5	9.25	0.77	0.0832	Straight	400	C3=/C3	0.35	0.050		1.84	4.3900	25.3	0.0625	3.081
7123-7130	2	37	24	0.5	9.25	0.77	0.0832	Straight	500	C3=/C3	0.28	0.050		0.84	6.1000	22.9	0.0625	2.973
7148-7155	0	37	24	0.5	9.25	0.77	0.0832	Straight	400	IC4/NC4	0.5	0.050		1.03	5.2900	24	0.0625	2.919
7158-7163	0	37	24	0.5	9.25	0.77	0.0832	Straight	500	IC4/NC4	0.45	0.050		0.31	7.9100	20.5	0.0625	2.757
7679-7684	1	37	24	0.25	9.25	1.19	0.1286	Straight	24	C6/C7	0.85	0.100		15.04	0.3076	42	0.0625	2.338
8065-8071	2	47	24	0.5	6.06	0.46	0.0759	Sloped	165	IC4/NC4	0.5	0.200		5.14	1.7800	30.7	0.0625	1.247
8075-8081	2	47	24	0.5	6.06	0.46	0.0759	Sloped	300	IC4/NC4	0.55	0.100		2.29	3.4600	26.9	0.0625	1.666
8091-8096	2	47	24	0.5	6.06	0.46	0.0759	Sloped	400	IC4/NC4	0.5	0.250		1.01	5.2100	24.1	0.0625	3.191
8109-8113	2	47	24	0.5	6.06	0.46	0.0759	Sloped	500	IC4/NC4	0.45	0.100	Unclear Efficiency Drop	0.24	7.9300	20.5	0.0625	2.574
8117-8123	2	47	24	0.5	6.25	0.46	0.0736	Sloped	24	C6/C7	0.75	0.150		14.56	0.3028	41.3	0.0625	0.434
8466-8470	1	30	24	0.1875	11.2	1.42	0.1268	Straight	1.933	IPA/H2O	0.75	0.150		20.77	0.0170	50.1	0.0625	0.457
8544-8549	1	30	24	0.1875	11.2	1.42	0.1268	Straight	0.309	O-P XYLENE	0.75	0.250		26.81	0.0064	52.9	0.0625	0.290
8562-8567	1	30	24	0.5	11.2	1.45	0.1295	Straight	0.193	C8/C10	0.9	0.300		19.28	0.0040	48.4	0.0625	0.190
9299-9307	1	30	36	0.5	11.2	1.18	0.1054	Straight	0.309	O-P XYLENE	0.85	0.150		26.66	0.0073	52.9	0.0625	0.193
9497-9503	2	47.4	24	0.5	5.13	0.439	0.0856	Sloped	165	IC4/NC4	0.6	0.200		5.14	1.7700	30.7	0.0625	1.236
9520-9526	2	47.4	24	0.5	5.13	0.439	0.0856	Sloped	300	IC4/NC4	0.5	0.150		2.39	3.3600	27.1	0.0625	1.586
9551-9556	2	47.4	24	0.5	5.13	0.439	0.0856	Sloped	24	C6/C7	0.9	0.500		14.7	0.3059	41.6	0.0625	0.762
9569-9575	2	47.4	24	0.5	5.13	0.439	0.0856	Stepped	165	IC4/NC4	0.6	0.200		5.1	1.7900	30.7	0.0625	1.228
9629-9634	2	47.4	12	0.5	5.13	0.439	0.0856	Sloped	165	IC4/NC4	0.75	0.250		5.13	1.7800	30.7	0.0625	1.578
9648-9656	2	75	24	0.5	37.2	2.96	0.0796	Semi-Slope	24	C6/C7	1	0.200	1-Pass 8' tower	14.9	0.3047	41.9	0.0625	3.853
9792-9798	2	75	24	0.5	37.2	2.96	0.0796	Semi-Slope	14.7	IPA/H2O	0.9	0.300	1-Pass 8' tower	16.89	0.1190	46.6	0.0625	1.607
9805-9812	2	75	24	0.5	37.2	2.95	0.0793	Semi-Slope	24	C6/C7	1	0.300	1-Pass 8' tower	15.14	0.2936	42.2	0.0625	1.853
9824-9831	2	75	24	0.5	37.2	2.95	0.0793	Semi-Slope	24	C6/TOL	1	0.100	1-Pass 8' tower	16.41	0.2740	44.2	0.0625	2.307
10461-10466	1	30	24	0.5	11.2	1.95	0.1741	Straight	0.387	O-P XYLENE	0.75	0.050		26.66	0.0071	52.9	0.0625	0.233
10484-10491	4	30	24	0.5	11.2	1.45	0.1295	Straight	0.387	O-P XYLENE	0.65	0.070		26.38	0.0078	52.8	0.0625	0.290
10508-10515	0	30	24	0.5	11.2	1.45	0.1295	Straight	0.387	O-P XYLENE	1.4	0.200		25.66	0.0108	52.4	0.0625	0.240
10536-10542	2.2	75	18	0.5	37.2	2.96	0.0796	Straight	14.7	C6/TOL	0.7	0.050	Clear DC	18.06	0.1880	45.2	0.0625	1.347
10554-10561	2.2	75	18	0.5	37.2	2.96	0.0796	Straight	24	C6/C7	0.7	0.050	Clear DC	14.46	0.3174	41.4	0.0625	1.880
13396-13400	2	37	24	0.5	9.25	0.77	0.0832	Straight	24	C6/C7	0.7	0.150	Weeping Study	13.37	0.3125	39.9	0.0625	0.916
14115-14119	2	37	24	0.5	9.25	1.27	0.1373	Straight	165	IC4/NC4	0.35	0.050	Weeping Study	5.01	1.8000	30.5	0.0625	2.495

Appendix B – Example Detailed Sieve Tray Stability Calculation for Total Reflux Runs 14115 to 14119

FRI Run No.	Liq Density	Vap Density	Flow Rate	Efficiency	Weir Load	Dry Drop	Ub	eta	Froth Height	alpha	Cd	HS	Stability
	lb/ft3	lb/ft3	lb/hr	%	gpm/inch	Inches of Liq	ft/sec		inches			Inches of Liq	
14115	30.6	1.7640	59170	113.0	6.52	1.373	1.007	3.761	7.83	0.2100	0.8433	1.645	0.913
14116	30.6	1.7680	46770	112.0	5.16	0.856	0.794	3.130	6.71	0.2422	0.7982	1.624	0.726
14117	30.5	1.8000	31580	103.0	3.49	0.384	0.527	2.284	5.27	0.3045	0.7409	1.605	0.489
14118	30.5	1.8000	22580	95.0	2.49	0.197	0.377	1.749	4.40	0.3638	0.7060	1.600	0.350
14119	30.3	1.8840	13870	62.0	1.54	0.071	0.221	1.157	3.53	0.4637	0.6713	1.638	0.209

Tower Diameter, inches	48
Bubbling Area, ft2	9.25
Top Downcomer Area, ft2	1.5
Btm Downcomer Area, ft2	1.5
Open Area, ft2	1.27
Hole Diameter, inches	0.5
Hole Pitch, inches	1.1875
Tray Thickness, inches	0.0625
Outlet Weir Height, inches	2.0
Outlet Weir Length, inches	37.0

619660 Eliminate Condensate Bottlenecks in Reboiler Discharge

Abstract:

Reboilers using an inlet steam control valve arrangement can provide significant economics and control benefits to certain distillation tower installations. By exposing the full surface area of a reboiler to steam for effective heat transfer, the steam pressure can be lowered - which can reduce the fouling that can occur with higher pressure steam and the corrosion that can occur with flooding a tube bundle. Control can improve because steam can adjust to process control demand changes much more rapidly than by the alternative of varying a liquid level - such as occurs with outlet valve control. Furthermore, it is generally not possible to lose the condensate seal/reboiler duty when the proper drainage design is implemented. Other potential benefits include mitigation of the stratification that can plague channel head gasketing in horizontal designs, and the use of "long" low pressure steam to improve the steam balance or reduce generation requirements.

However, in order to achieve those benefits, inlet steam control has to overcome certain potential operational pitfalls, which can create bottlenecks in the system drainage and hinder process operation. These include insufficient pressure to discharge condensate into the return header on start-up, under low demand, or when the tube bundle is relatively clean. Also, there can be design uncertainty relative to the drainage system utilized – steam trap (and if so, which type?), pump/trap, level pot, electric pump? Moreover, insufficient understanding of balance requirements for level pots or pump/trap reservoirs can lead to reboiler evacuation issues, such as stall, backup, hammer, and corrosion.

To help improve reboiler performance when using inlet steam control valve operation, this presentation reviews historical negative experiences and provides insight into design options to mitigate condensate backup in reboiler operation

620065 Designing Condensate Drainage from Reboilers

Abstract:

Reboilers using an inlet steam control valve arrangement can provide significant economic and control benefits to certain distillation tower installations. However, their reliability requires effective condensate drainage from the steam space. Unfortunately, there are multiple instances of reboilers experiencing water hammer-related issues such as channel head gasket and tube bundle damage or the sewering of condensate. These issues may be more commonly experienced with horizontal type reboilers.

Many of these problems can be avoided/mitigated with a proper condensate drainage design, but it often occurs that the reboiler system experiencing difficulty had improper original design or installation and requires correction for improvement. **This presentation explores why certain problems occur and provides engineers with some specific key design considerations necessary for proper performance in draining reboilers.**

619671 Reuse and Relocation: A Case History Part II reboiler Analysis and Performance

Abstract:

Potentially, used equipment can save time and money over new equipment in your plant. Occasionally, used equipment can be picked up for very low prices. But getting it to the plant and properly working does require skills that differ from purchasing new equipment.

However, the process requirements for the new service are nearly always different from the original process requirements. Expert-level knowledge may be required to decide if the equipment will really do the required job.

Unless an entire plant is purchased and used as-is, used equipment will have to be integrated with other parts of the plant that will use new equipment. Integration may be difficult as the used equipment may have constraints that makes its use difficult with conventional choices for the new equipment attached to it.

This presentation examines a plant that purchased a 42-inch diameter tower used from a pharmaceutical plant. It had been built and erected but never placed into service. The client purchased the tower and placed it into a completely new service.

The focus on this presentation is on the reboiler system and difficulties with this system. The reboiler chosen was a falling-film evaporator. Once in service, the reboiler could not get the required duty into the tower. Areas covered include:

- Investigating leaks in the reboiler,
- Determining the actual reboiler duty,
- Examining the actual capability of the hot oil system providing heat to the reboiler,
- Inherent limits on the reboiler performance including fouling and liquid distribution problems,
- Steps taken to improve reboiler performance, and
- The alternate best case reboiler type that would have performed better.

Finding the Proverbial "Needle In The Haystack"

Derrick K. Rollins, Sr.[1,2]
Iowa State University, Ames, Iowa 50011

[1]Department of Chemical and Biological Engineering
[2]Department of Statistics

Often the tasks of Big Data Analysis and/or Discovery are impeded by the size of data sets. Moreover, when data sets are large and the desired information content in the data is quite small, the ability to discover and/or analysis this content is extremely challenged. This is akin to a real mining effort when most of the solids and elements are not the precious "gems" being sought.

Wang et al. (2020) obtained GSE994 from the Gene Expression Omnibus database. This data base is composed of gene expression data taken from lung tissue samples of former, current, and never smokers. The goal of the study was to elucidate the relationship between cigarette smoking and ACE2 expression. ACE2 is the primary mode of entry for the SARS-CoV-2 virus that is responsible for causing COVID-19. More ACE2 receptors provides the virus with more opportunities to infect someone. Wang et al. created One Pseudo Variable that was some kind of correlation for each gene and kept "the significant ones" and created **one** pseudo variable. They used this pseudo variable to obtain a score for each person that they called "relative ACE2 expression level."

However, there are several critical limitations of their approach such as: 1. only one pseudo variable; 2. analysis based only on the three reported classes (Current, Former, Never); 3. no potential for subclass analysis and thus; 4. deep data mining is not possible.

In contrast to the Wang et al. approach, this talk presents a powerful multivariate statistical approach based in Principal Component Analysis (PCA). However, it is a unique application of PCA, as it does not keep the top most latent variables that explain a certain threshold of the variation and discard all the rest. This typical approach can miss rare "gems" that show up in the extreme low percentages of the variation. The talk presents a PCA methodology that can find critical behavior (i.e., effects) in very large data sets when the number of experiments are very small in comparison to the number of features for each experiment. It exploits a methodology developed by Rollins et al. (2006) that uses what they termed as Eigen Genes (EGs) and Eigen Assays (EAs). These are the pseudo variables that are used by this approach. An assay is an experimental sample or feature and in the smoker study here, they are the smoking history of the 75 people in the study. Thus, there are 75 assays giving 75 EGs and 75 EAs or 150 pseudo variables for the Rollins et al. approach. All of them are analyzed looking for subgroups within the specified three groups that have statistically significant behavior. This deep analysis identifies subgroups with as many as one subject.

The overall conclusion of this study is that the most powerful method is the one that can most accurately "score" each assay based on its true nature and can do this with high sensitivity even when it is a fraction of the information in the data set. Additional conclusions are:

- Methods that are not PCA based are very limited in analyzing Gene Expression Data
  - The Use of EGs and EAs to Find Physiological and Biology Groups has Both Depth and Breath that is Quite Unmatched
    - Groups are formed and analyzed based on the data
    - This makes it practical and possible to find those "needles in the haystack"
- Simultaneous Confidence Intervals are More Informative than P-values, Because, in Addition to Reliability, They Show Precision
- Effective Methods Should Have Strong Agreement in the Ranking Genes

Literature Cited:

Jin Wang, Qiulin Luo, Rui Chen, Tao Chen and Jian-xiang Li, "Susceptibility Analysis of COVID-19 in Smokers Based on ACE2," doi:10.20944/preprints202003.0078.v1 (2020).

Rollins, D. K, D. Zhai, A. L. Joe, J. W. Guidarelli, and R. Gonzalez, "A Novel Data Mining Method to Identify Assay-Specific Signatures in Functional Genomic Studies," BMC Bioinformatics, 7 377 (2006).

620773 Hydrocracker work-up section revamp with improved ConSep trays

Abstract:

During 1999 a revamp of a hydrocracker work-up section was executed which included the use of ultra-high capacity ConSep trays. 20 years later this work-up section has again been revamped targeting a significantly higher conversion. The presentation will deal with the modifications made in the hydrocracker work-up section scheme and contain details about improvements made to ConSep trays in the last decade.

620482 Recovery and Stripping issues at turndown conditions

Abstract:

During the refinery cutbacks in the first part of 2020, A FCC unit experienced LPG recovery and H_2S stripping issues. The H_2S in the naphtha was extremely high and could not be influenced by adjustments in the Stripper/Absorber or Debutanizer. The issue was isolated to one unit during a time when all the FCC's were turning down due to reduced demand.

The troubleshooting was extensive and involved analytical, equipment rating and feed quality evaluations. The process manipulations were extensive and provided little benefit. The final solution turned out to be very simple and unexpected.

620483 Limit/Prevent Downtime by Designing Atmospheric/Vacuum Tower Lines for Upset Condition

Abstract:

Continuous Distillation throughput can be limited by operational issues with product solidification (plugging) in interconnecting and downstream piping and equipment. These events, while seldom, have historically incurred high cost and lengthy timeframes to rectify – reducing unit profitability.

This presentation will examine specific operational upsets and the solutions plants implemented to eliminate or shorten future downtime occurrences due to plugging in tower bottoms and downstream lines . We'll also discuss general heating system design objectives that consider upset conditions to prevent issues and guard against and/or allow for quicker, more cost-effective recovery from unexpected downtime.

The Emergence and Evolving Role of Big Data in Polyethylene Resin Development

David M. Fiscus

Big data knits together and provides objective understandings of the interplay and importance of many variables, like production, material, performance and market variables. It identifies how to effectively supply key attributes, while improving profitability and efficiencies in developing new materials / products. It is being used to develop new theories and methods for improving material performance; for example, relating resin molecular architecture to microstructure and micromechanical deformations. This talk will discuss the evolving role of Big Data in R&D product development over a 30-year career working with polyethylene resins. It will highlight challenges encountered, milestones achieved, and future possibilities

AUTHOR INDEX

AIChE
120 Wall Street, FL 23
New York, NY 10005-4020

ISBN 978-1-7138-5333-6

CURRAN ASSOCIATES INC.
proceedings.com

9 781713 853336